# Concepts & Calculations in Analytical Chemistry
## A Spreadsheet Approach

## Henry Freiser

Department of Chemistry
University of Arizona
Tucson, Arizona

CRC Press
Boca Raton   Ann Arbor   London   Tokyo

**Library of Congress Cataloging-in-Publication Data**

Freiser, Henry, 1920–
    Concepts and calculations in analytical chemistry: a spreadsheet approach/Henry Freiser.
      p.  cm.
    Includes index.
    ISBN 0-8493-4717-3
    1. Chemistry, Analytic—Data processing. 2. Electronic spreadsheets. I. Title
QD75.4.E4F74   1992
543′ .00285—dc20

          92-3290
          CIP

Direct all inquiries to CRC Press, Inc., 2000 Corporate Blvd., N. W., Boca Raton, Florida, 33431.

International Standard Book Number 0-8493-4717-3

Library of Congress Card Number 92-3290
Printed in the United States    2  3  4  5  6  7  8  9  0
Printed on acid-free paper

# Errata

## *Concepts & Calculations in Analytical Chemistry*

p.  17  Example 1.4 lines 8-9 should read $F(x) = A2\hat{}6 - 45.65*A2\hat{}5 + 794.47*A2\hat{}4 - 6666.67*A2\hat{}2 - 54737.74*A2 + 37942.7$

p.  21  Problem 1-1a (revised for those who cannot locate an old edition of the Handbook). Develop a spreadsheet for each of the reagents in the accompaning table in the manner of example 1.1.

### Specific Gravity of Aqueous Reagents at 20°C

| Weight % | Acetic Acid | HCl | $HNO_3$ | $H_3PO_4$ | $H_2SO_4$ | $NH_3$ |
|---|---|---|---|---|---|---|
| 1 | 0.9996 | 1.0032 | 1.0036 | 1.0038 | 1.0051 | 0.9939 |
| 2 | 1.0012 | 1.0082 | 1.0091 | 1.0092 | 1.0118 | 0.9895 |
| 3 | 1.0025 | | 1.0146 | | 1.0184 | |
| 4 | 1.0040 | 1.0181 | 1.0201 | 1.0200 | 1.0250 | 0.9811 |
| 5 | 1.0055 | | 1.0256 | | 1.0317 | |
| 6 | 1.0069 | 1.0279 | 1.0312 | 1.0309 | 1.0385 | 0.9730 |
| 7 | 1.0083 | | 1.0369 | | 1.0453 | |
| 8 | 1.0097 | 1.0376 | 1.0427 | 1.0420 | 1.0522 | 0.9651 |
| 9 | 1.0111 | | 1.0485 | | 1.0591 | |
| 10 | 1.0125 | 1.0474 | 1.0543 | 1.0532 | 1.0661 | 0.9575 |
| 11 | 1.0139 | | 1.0602 | | 1.0731 | |
| 12 | 1.0154 | 1.0661 | 1.0661 | 1.0647 | 1.0802 | 0.9501 |
| 13 | 1.0168 | | 1.0721 | | 1.0874 | |
| 14 | 1.0182 | 1.0675 | 1.0781 | 1.0764 | 1.0974 | 0.9430 |
| 15 | 1.0195 | | 1.0842 | | 1.1020 | |
| 16 | 1.0209 | 1.0776 | 1.0903 | 1.0884 | 1.1094 | 0.9362 |
| 17 | 1.0223 | | 1.0964 | | 1.1168 | |
| 18 | 1.0236 | 1.0878 | 1.1026 | 1.1008 | 1.1243 | 0.9295 |
| 19 | 1.0250 | | 1.1088 | | 1.1318 | |
| 20 | 1.0263 | 1.0980 | 1.1150 | 1.1134 | 1.1394 | 0.9229 |
| 100 | 1.0498 | 1.1980 (40 wt. %) | 1.5129 | 1.870 | 1.8305 | 0.8920 (30 wt. %) |

p.  39  Please correct my omission of the factor 1/2 from the right-hand side of Equations 3-4 and the unnumbered equation at bottom of p. 39.

p.  40  The correct equation for $K_2SO_4$ is $I = 1/2(0.1 \times 2 \times 1 + 0.1 \times 2^2) = 0.3$

P. 145  The equations defining the numerators of the **f** fractions of the several U species were mistakenly transposed. The numerator of $f_3$ should be $K_{65}K_{54}K_{43}$, that for $f_4$, $K_{65}K_{54}\pi$; $K_{65}\pi^2$, that for $f_5$, and $\pi^3$, the numerator for $f_6$. The correct formulation is shown here.

$$C_U = [U(III)] \left\{ 1 + \frac{\pi}{K_{43}} + \frac{\pi^2}{K_{54}K_{43}} + \frac{\pi^3}{K_{65}K_{54}K_{43}} \right\}$$

OR

$$f_3 = \frac{[U(III)]}{C_U} = \frac{\pi_3}{\pi^3 + K_{65}\pi^2 + K_{65}K_{54}\pi + K_{65}K_{54}K_{43}} = \frac{K_{65}K_{54}K_{43}}{D}$$

$$f_4 = \frac{K_{65}K_{54}\pi}{D} \quad f_5 = \frac{K_{65}\pi^2}{D} \quad f_6 = \frac{\pi^3}{D}$$

Please note that the general rules listed on p. 145 as (a), (b), and (c) are correct as written. In using the analogy between the $f$ values of redox systems and the $\alpha$ values in acid-base systems, think of the highest oxidation state, whose numerator is $\pi^n$, as analogous to the fully protonated species, whose numerator is $[H^+]^n$.

p. 185      In Table 9.1, the log $\beta_{MIn}$ values for EBT should read: 15.0 (Zn), 12.6(Mn), ~20(Cu). Values for Mg and Ca are correct. For Calmagite, correct log $\beta$ for Mg to 8.9. For Pyrocatechol Violet, add the following log $\beta$ values: 16.5(Cu), 10.4(Zn), 9.4(Ni). MIn for PV are blue.

p. 187      On the first line, log $\beta K'_{MgIn}$ should read log $\beta'_{MgIn}$

p. 190      Equation (10-1) should read: $V_R[Ox]_A n_A = V_o[Red]_B n_B$

p. 191      Delete Equation (10-1a). Change Equation (10-3) to read (substituting in 10-1):

$$n_A V_R C_R f_{Ox(A)} = n_B V_o C_o f_{Red(B)}$$

and finally,

$$V_o = \frac{V_R C_R n_A}{C_o n_B} \cdot \frac{f_{Ox(A)}}{f_{Ox(B)}}$$

p. 193      The tabulated values in Table 10.1 are the $E°$ not the $pE°$ values. The corresponding $pE°$ values are 18.75, 21.11, 14.36, 8.95, 18.92, and 21.79.

p. 195    Substitute the following problem for the present 10-2, which is too involved at this point because the ratio of [Ox]/[Red]² rather than [Ox]/[Red] appears in the Nernst Equation, resulting in a titration curve that changes with the absolute concentration:

**Problem 10-2**

Develop the curve for the titration of 50 mL of 0.2 M SnCl₂ with 0.05 M KMnO₄ in 1.0 M HCl (assume pH = 0). We will ignore the possibility of the oxidation of Cl⁻.

p. 242    The entire page up to Example 16.3. The treatment given is not clear. Please replace by:

A value of $D_K$ = 105.2 gives the %K⁺ exchanged as 10 × 105.2/(105.2 + 50) = 67.7%, signifying that the assumption of total exchange is incorrect. To obtain a correct solution, let x = mmols K⁺ exchanged, then D = [K⁺]$_r$/[K⁺] = x/[(0.2 − x)/50] = 50x/(0.2 − x). D is also K$_{ex}$[H⁺]$_r$/[H⁺]. Hence,

$$\frac{50x}{(0.2-x)} = \frac{2.28(5-x)}{0.1+\dfrac{x}{50}}$$

This results in a quadratic equation which readily yields the result of x = 0.1366 and D = 107.7. The equation can also be solved by using the Pointer function after transposing one side to the other and the entire expression, F(x), to get **@log(@abs(F(x))**.

When multicharged ions are exchanged the D values, and therefore the % ion exchanged, are usually large enough so that reliable results are obtained when total exchange is assumed.

vith Chapter 1, spreadsheet techniques are introduced and developed as
1e nature of the problems and examples described. A very minimum of
1eeds to be reviewed in detail. The student is urged to explore the many
1enus which are self-explanatory. The text develops a basic spreadsheet
:h includes the following.

## Data and Formulas for Multiple Calculations

a spreadsheet problem by entering labels for $i$, the critical variable and for
her variables derived from the critical variable, in the first row of as many
needed. Next, by filling (/**EF**;/BF;/DF)[2] column **A** with a desired range
f the critical variable in specified intervals. With over 8000 rows available
vals can be as small as anyone would need. Examples in the text may call
)00 rows, but 100-200 rows are common. Then, in the succeeding columns
he **top row only**, the algorithms for all of the functions derived from the
ue as needed for the problem. All of the rows can be filled by simply using
ommand (/EC;/BC;/C). The student focuses attention on the principles and
arithmetic!
entering parameters such as stoichiometric concentrations and equilibrium
use of **Absolute Addresses**, or complex problems, **Block Names** permits
ve many problems of the same type without writing another spreadsheet,
y by entering the new parametric values.

## 1 of Complex Equations

eadsheet solution of the method of **Successive Approximations** which all
inates the need for the traditional Newton-Raphson method as well as the
tion of a **Pointer Function** defined as **@LOG(ABS(F(x)))** which, in the
lisplay of the system, has a distinct minimum pointing at the answer to the

## Fitting: Regression Analysis

spreadsheet's Regression function (/**TAR**,/AR,/DR) renders very simple the
of data such as a spectrophotometric calibration curve by linear regression
. Data from a multicomponent unknown having characteristic spectra for

nands for filling a block of cells with a series of regularly spaced numerical values. Note that the
dsheet commands in this text are given first in QPro, then, in brackets, the earlier but widely used
:ro and Lotus 123. It is a useful exercise for instructor and students to examine each of these menu
to become acquainted with QPro, Q, and 123. They may be called from QPro by the command
1, from Q by /DSM, and from 123 by /WGDFM. It will be noticed that the differences in
nand are more apparent than real. Particularly helpful is the brief explanation that appears at the
m left of the screen for each command.

# FOREWOR

This text represents an effort to organize the cor
cal chemistry into a systematic unified whole in whi
area strikes resonances helpful in the mastery of (
between simple and advanced treatment is, in ma
conceptual, the integration of computer spreadshee
destroys this barrier to the advantage of students w
richer picture of the subject.

Two principal features characterize the systematic
lations used in this book: a) Expressing concentra
product of $\alpha$ (a fraction of that species of all other:
fractions are a function of only the critical variable ((
rium constants, and C, the total concentration of the (
the equilibrium condition by a single 'balance' equa
equation (PBE), the ligand balance equation, etc. This
tion of the equilibrium condition of the solution by
concentration varible, i.e., in an implicit solution.

A third important feature of this text is the thorough
techniques not only to eliminate the arithmetic barri
graphical visualization of entire problem 'surfaces'. It ca
this advance does not require knowledge of computer p
With a spreadsheet, the user is totally free, using a trul;
to accomplish wonders of his own. The student will rejc
the spreadsheet to suit individual problems (simplicity its
student will see major improvement in his grasp of fundar

This approach is effective not only for calculations invc
The PBE and other balance equations apply to titration c
A single, easily derived equation suffices to describe all
be it for acid-base, complexometric, redox, or precipitatic

## Role of Spreadsheet in Simplification and Visuali

What a glorious opportunity awaits us when we discov
sheets! The simple elegance of graphical solutions of num
briefly somewhere around the 7th to 9th grade in public
solution of simultaneous linear equations". It is puzzling a
most of us, it then sinks without a trace while we diligent
ways. The obvious pedagogic advantages of viewing proble
merely settling for one specific answer have been largely ov

[1] Calculations for individual points on titration curves are also treated in th

Starting
needed by
commands
pull-down
toolkit wh

## Entering

Starting
ii, all the o
columns a
of values
these inte
for up to
write, in
critical v
the copy
not on th
When
constant
one to s
but mere

## Solutio

A sp
but elin
introduc
graphic
problen

## Curve

The
analysi
analys

2   Con
spre
Qua
typ
/OS
con
bot

each component can be analyzed by multiple regression analysis to give the composition and standard deviations of the composition of the unknown.

## Calculus Functions

Because values of the critical variable can be listed in very small increments without the customary tediousness, the spreadsheet format is very well suited for performing reasonably accurate calculus operations by numerical differentiation and integration. Useful examples of the former include titration curve slopes, $dV_b/dpH$ and $dpH/dV_b$, which give us important titration curve parameters, namely, the buffer and sharpness indices. Another area of great interest, chemical kinetics, represents an additional topic where numerical differentiation and integration is of great use.

Numerical integration is used in the text to show how confidence limits can be derived from the Gaussian error curve. It also enables us to obtain concentrations by measuring areas under chromatographic bands. This procedure is also useful in the analysis of spectral absorption bands.

## Graphics

Graphical representation is probably the most dramatic contribution to the full understanding of concepts and calculations because of the global view this affords of the problem. There are over 60 graphs developed in the text to illustrate this vital principle.

With QPro, the Annotate feature allows writing of text and symbols on the graphs. Additionally, this function helps overcome the annoying limitation of these spreadsheet graphs to a maximum of 6 individual curves by allowing you to use it for linear curves without using one of the 6. In all three of the spreadsheets mentioned, the scale function for both X and Y provide a ZOOM feature of great utility in increasing definition of an answer or a part of the graph.

## Utilization of Error Curve in Statistics, Separations, and Spectrophotometry

Numerical calculations involving the Error curve are certainly made very easy by means of the spreadsheet. This applies not only to statistical considerations, but to simulation of chromatograms as well. Explanation of band shapes is elucidated by attributing their characteristics to those of the Gaussian curve. Thus, bandwidth is related to $\sigma$, the standard deviation, $V_r$, the retention volume to $\mu$, the mean, as is well known in contemporary discussion of chromatography. Similarly, one can simulate the spectra of a multicomponent mixture by assuming each spectrum can be represented by a Gaussian curve with $\mu$ a model of $\lambda_{max}$ and $\sigma$ chosen to match a reasonable spectral band width.

This text could serve as a model for other chemistry courses as well as for other scientific, engineering, and mathematical courses. I have found the ease with which even students near the beginning of their training can cope with complex calculations and sophisticated concepts to be most impressive. So, too, is the way in which the student can produce and digest large assemblies of numercial data/calculations while still focusing mainly on the underlying principles governing the topic. A further advantage derives from the inherent attraction that almost anything connected with computers has for students (and teachers, too).

The program used to generate the spreadsheets and figures illustrated in this book is QuattroPro(QPro). QPro is a powerful and versatile spreadsheet available in an academic version at a reasonable price, ~$50. Several QPro features are of immediate and practical utility for the classroom. A series of figures, text messages, and other graphics can be displayed as a Slide Show. Also, figures may be annotated (e.g., arrows, lines, and labels for individual parts of the diagrams).

# ACKNOWLEDGMENT

I am deeply obligated to my many teachers, colleagues, and, most of all, students, whose teachings and criticisms have over the past half century helped shape and enrich my understanding of the science of analytical chemistry. With respect to this book, I would like to acknowledge the major role played by my research group, particularly Dr. Roger Sperline, for introducing me to the use of spreadsheets. I am grateful for the tolerance of my graduate and undergraduate classroom students on whom I tried some portions of this text. I deeply appreciate Dr. Ben S. Freiser's careful reading and helpful suggestions of much of the material. Most of all, I affectionately acknowledge the help, support, and encouragement of Edie, whose 50th wedding anniversary has quite suddenly come up.

# A SPECIAL WORD TO STUDENTS

A major objective of this book is to help you become familiar with and confident about the important ideas used by analytical chemists. These ideas are also essential to others who make physical and chemical measurements the basis for their decision-making. This includes chemists in general, engineers, physicists, physicians, etc. In it we will examine the ideas themselves and their relation to each other. By this means, we will develop appropriate, rigorous equations which will be made "user friendly" by the techniques used here to solve them. Together we will conquer all tedious and repetitive arithmetic operations with the help of a computer spreadsheet program. The one we will use is called QuattroPro(QPro). We will give references in the text to the QPro menu and commands but equivalent menus, such as those for Lotus 123, or the older Quattro version can be called up.

This book is not a spreadsheet manual, however. It is an analytical chemistry text which derives benefit from the liberal use of the spreadsheet approach. What you learn here will have a bearing in problem solving not only in other chemistry courses but in most of your other science, engineering, and mathematics courses, as well.

I also want to mention the frequent use in this book of graphs to represent and clarify problem situations. Combining graphical and algebraic approaches to problem solving is a relatively neglected but delightful way to eliminate most of the work of analyzing masses of data. Problems will be made more orderly, and analyzed much more readily, by means of visual relationships. The added benefit of the spreadsheet is the utter simplicity of generating all manner of graphical displays. This will be fully exploited in this book. Even in our highly developed technological society, however, you cannot always count on having a computer handy when you need one. Therefore, this text will also use the combined graphical-algebraic approach utilized in earlier work. This has as its objective the mastery of a systematic method of simplifying complex algebraic expressions until they can be solved both simply and with reasonable accuracy ($\pm$ 5%).

I know that many of you are already familiar and comfortable with the personal computer and its magic ways. Let me say to those who aren't that I am a recent learner myself. Using a spreadsheet program has enabled me to overcome totally my earlier feeling that computer programming was something to be left to "computer jocks". Those who developed spreadsheets must have had us in mind. Their genius has resulted in a product that is very easy to learn and to use. With the aid of a spreadsheet you still do not have to program, but you can get to where you are going simply by following menu-driven instructions. Just wait and see for yourself!

If you are like the students with whom I have worked this way, I fully expect you to be able to help your instructors with spreadsheet techniques before too long. Be gentle with them or, at least, subtle.

Henry Freiser
Tucson, Arizona
January, 1992

# TABLE OF CONTENTS

# Chapter 1

# Introduction

---

Poised as we are on the threshold of the 21st century, we can still derive help and inspiration from prior scientific thinkers of the 18th and 19th century as well as from our own. It has long been an accepted axiom that "mathematics is the handmaiden of the sciences." In just the same way, analytical chemistry is the handmaiden of the experimental sciences. As Professor I. M. Kolthoff said long ago, "Theory guides, but experiment decides." Analytical chemistry, in its broader sense, the science of measurements, is the key, the emblem of the way in which a scientist, standing on the threshold of an important hypothesis, can pass through to its validation. Analytical chemistry provides access to the testing and validation of chemical hypotheses.

The science of chemistry arose out of alchemy when quantitative measurements began to be employed to characterize the course of chemical reactions. Chemists are primarily interested in elements, compounds both pure and in mixtures as well as in their chemical transformations. We link the counting of atoms and molecules to more accessible properties of matter which are related to number, such as mass, volume, and optical absorbance, to name a few. A beautiful structure of chemical principles has been developed in the past two centuries which enables us to make maximum use of measured quantities for the purposes of counting atoms and molecules and thereby to characterize chemical substances and chemical reactions. Also of central interest to chemists are the great strides that have been made in our century in overcoming earlier limitations of observations and computation. The pulse of the computer is far more rapid than ours; when we do something in a twinkling, we mean a fraction (0.2-0.3 second); the computer's "twinkling" it requires to perform an operation is a microsecond and is well on the way to being a femtosecond. This has a profound and direct impact not only on experiment but on chemical calculations as well.

## STOICHIOMETRIC CONCEPTS

A review of elementary chemical calculations naturally starts with various means of describing, in "chemical terms", i.e., in terms of numbers of molecules, amounts and concentrations of substances that take part in chemical reactions.

## The Mole Concept

From a chemical viewpoint, there can be no more important characteristic of a sample of matter than the number of chemical units or molecules it contains. Although other properties of the sample such as mass, volume, etc. are useful, we must relate these to the number of molecules in a substance in order to predict easily, from the balanced chemical equation, what quantities of the substances in the same reaction will be required or produced.

To be sure, working with the actual numbers of molecules is awkward because there are so many of them in even the minutest portions of matter. This awkwardness can be removed, however, by using $6.023 \times 10^{23}$ molecules (Avogadro's number), the chemist's "dozen", as a basic unit. The amount of matter containing an Avogadro number of molecules is called a **mole**. This amount will have a mass in grams equal to its gram molecular mass (GMW). To calculate the number of moles of a substance from its mass in grams,

$$\text{Number of moles} = W/GMW \qquad (1\text{-}1)$$

For instance, if we were considering a reaction in which NaOH neutralizes $H_3PO_4$, from the balanced equation:

$$H_3PO_4 + 3NaOH \rightarrow Na_3PO_4 + 3H_2O$$

we see that if $x$ moles of $H_3PO_4$ are used, $3x$ moles of NaOH are required for neutralization. This produces $x$ moles $Na_3PO_4$ and $3x$ moles of $H_2O$. A consequence of these simple definitions allows us to write expressions for the amounts of substances related by participation in the same (set of) reaction(s). Let us consider the question of how much $BaSO_4$, $xg$, can be produced from 1.00 g of NaCl by the following reactions:

$$2NaCl + H_2SO_4 \rightarrow Na_2SO_4 + HCl\uparrow$$
$$Na_2SO_4 + BaCl_2 \rightarrow 2NaCl + BaSO_4\downarrow$$

From the balanced equations, the number of moles of $BaSO_4$ equals the number of moles of $Na_2SO_4$, which in turn is equal to half the number of moles of NaCl, or

$$\text{moles BaSO}_4 = \text{moles Na}_2\text{SO}_4 = 1/2 \text{ moles NaCl}$$
$$x/\text{GMW BaSO}_4 = 1/2 \cdot 1.00/\text{GMW NaCl}$$

Rearranging,

$$\text{x gBaSO}_4 = 1.00 \text{ gNaCl} \cdot (\text{GMW BaSO}_4/2\text{GMW NaCl})$$

The expression in parentheses, the **corrected** ratio of molecular weights of the substance sought to that of the given substance is commonly used in gravimetric analysis and is called the **gravimetric factor.** This is very useful in relating amounts of reacting substances and products involved in the same set of chemical reactions. The weight of any desired reactant or product that is obtainable from a given weight of a particular substance involved in a series of (or only one) reaction(s) is simply the product of the weight of the given substance and the appropriate gravimetric factor.

## Expressions of Concentration

Many of the chemical reactions of interest will involve solutions rather than pure substances. Hence, to find the amount of substances in solution, we must know its concentration (amount/volume) as well as the solution volume. Of the means of expressing concentration, molarity (M), molality (m), and mole fraction (x), are the most important.

## Molarity

Molarity, M, is expressed in terms of moles of solute per liter of solution. If 11.10 g $CaCl_2$ (whose GMW = 110.99 g/mol) is dissolved in enough water to make 0.500 L of solution, its molarity is seen to be

$$M = 11.10 \text{ g}/110.99 \text{ gCaCl}_2/\text{mol}/0.500 \text{ L} = 0.200 \text{ mol/L}$$

In general, M = number of moles per number of liters. This definition may be rearranged to give

$$\text{Number of moles} = M \times V \text{ (with V in liters)} \qquad (1\text{-}2)$$

Because the volume is usually measured in milliliters (mL), the unit millimole, mmole, is convenient. The molarity can be expressed as mmole-/mL, which will have the same numerical value as when it is in mole/L.

Hence the product

$$\text{Number of mmoles} = M \times V \text{ (with V in milliliters)} \qquad (1\text{-}3)$$

It is of the utmost importance to remember that moles represent an amount, whereas molarity signifies a concentration, i.e., an amount per unit volume. For example, the molarity of either 1.0 mL or 10 L of 0.10 M HCl is the same; but the number of moles of HCl in these two solutions is vastly different.

A 0.1 M solution of NaCl or $CH_3COOH$, or any other substance, is a solution that is made by dissolving one tenth of a mole of the substance in water and making the volume of the solution up to a liter in water. This molarity describes the analytical (or total) concentration of the component, e.g., NaCl or $CH_3COOH$, rather than the concentrations of individual species at equilibrium. Some texts describe the analytical concentration by the term **formality**, F (i.e., the number of gram formula weights of solute per liter solution), and reserve the molarity for the equilibrium concentration of a given species.

In this text, whenever reference is made to an x M solution, it will mean that the solution we are describing is made by dissolving x moles of the substance in water and making the resultant solution up to a liter with water. Such total or analytical concentrations will be represented by the symbol C with a suitable subscript. Square brackets will be used only to denote the molar concentrations of the molecular or ionic species in the solution, e.g., [HOAc], [OAc⁻], or [Cl⁻].

The most convenient unit of concentration for use in equilibrium calculations is the molarity, and is the most frequently used concentration in this book. Using molarity has certain disadvantages, however. First, to calculate the exact amount of solvent present in the solution, it is necessary to obtain the density of the solution at the temperature desired. Since the density of the solution varies with temperature, the molarity will also vary. To obtain the water content in a 0.1000 M NaCl solution, for example, we need to know the density of the solution at 20° C (from the *Handbook of Chemistry and Physics* (CRC Press), P = 1.0028 g/mL). The mass of a liter of the solution is 1002.8 g. Of this, 5.845 g (corresponding to 0.1000 mole) is NaCl. Hence, 997.0g is the mass of water. Five degrees higher, at 25°C, this solution occupies a volume of 1.0018 L, but the amount of salt has not changed, which means that the molarity has dropped by 0.1%.

(With more concentrated solutions, this temperature effect is more pronounced.) For most purposes, however, such changes do not introduce any appreciable error in equilibrium calculations.

### Molality

Molality is a unit of concentration expressed as moles of solute per kilogram of solvent. This method of expressing concentration avoids the drawbacks seen in molarity. The strength of a solution expressed in molality is temperature independent and the amount of solvent easily calculable. This is achieved, however, at the expense of our knowledge of the exact volume of the solution, a distinct disadvantage in analytical operations involving volumetric ware. In such operations, use of molarity is preferred. In all equilibrium calculations, the advantage of having a concentration unit which is related to a definite composition makes molality the preferred choice.

Since most aqueous solutions described in this text are relatively dilute, the molarity is sufficiently close to the molality for it to be used in its place as a reasonable approximation. This is illustrated by calculating the molality of the NaCl solution described above:

$$m_{NaCl} = \text{moles NaCl/kg solvent} = 0.1000/997.0/1000 = 0.1003 \quad (1\text{-}4)$$

which is only a 0.3% difference from the value of the molarity of the solution. See Problem 1.1 for more details on the magnitude of the difference between M and m as a function of the concentration.

### Mole Fraction

Mole fraction is a unit of concentration expressed as the ratio of the number of moles of solute to the total number of moles of all components in the solution. In the case of the NaCl solution above,

$$X_{NaCl} = \frac{0.1003}{0.1003 + \dfrac{1000}{18.016}}$$

Or, in general

$$X_A = \frac{m_A}{m_A + \dfrac{1000}{(GMW)_{solvent}}} \qquad (1\text{-}5)$$

where $m_A$ is the moles of solute A, and GMW is the molecular weight of the solvent.

Mole fraction would be a desirable means of expressing concentrations for equilibrium purposes and, indeed, is often used for simple solutions containing only a few components. It is especially useful in comparing extraction distribution constants of a solute for different organic solvents.

## SPREADSHEETS IN CHEMICAL CALCULATIONS

The computer spreadsheet approach provides the means of carrying out a large number of calculations with the same ease as a single calculation. They can be used to solve highly complex equations by removing most of the work and by removing all of the tedium of the method of successive approximation. Even better, a novel approach, called the Pointer Function, leads to unique solutions of complex equations in a single step. The spreadsheet has useful database functions such as arranging information in a related series of "fields" which can be sorted according to a variety of selected logical rules. These procedures will be widely used throughout this text but will be illustrated here. You are bound to be impressed with the wealth of detail about the organization and capabilities of QPro that is revealed in the Manual. Just by using / followed by the initial letter of File, Edit, Style, Graph, Print, Database, Tools, Options, Window, you will see the respective 'pulldown' menus which begin to reveal this. Follow this up by reading the section in the QPro manual on Organization of the Spreadsheet. (If you are using Quattro or 123, carry out the corresponding menu examination.) It is important not to let this overwhelm you. There will be time to become familiar with all of this. At the outset, let us simply focus our attention on those relatively small sections that we need to know to do the problems at hand. Since spreadsheet manuals generally use business activities rather than science/engineering examples, let us take a moment to indicate the main ways in which we will be using this marvelous tool. In general, we will start by either (a) typing (or importing) data into the leftmost column(s), or (b) specifying a series of values (using /**EF**{/**BF**;

/DF})[1] of an independent variable located in column A. We can now enter relevant formulas expressing the functional relationships to the data from (a) or (b) in to calculate single values of quantities (dependent variables) in the first row in each of the columns needed. These values are repeated for all the dependent variables simply by keying in the copy command: /EC{/BC;/C}. Let us consider some examples of each of these categories here.

## ENTERING DATA AND FORMULAS FOR MULTIPLE CALCULATIONS

In the first example, chosen to show that it is possible to solve a whole family of problems in little more time than it takes to do a single one when one doesn't use the spreadsheet, we will learn how to **label** columns needed in the exercise, dealing with expressions of concentrations of aqueous hydrochloric acid, **enter data** of interest, and **use formulas** so that the data can be transformed into the desired answers to the question(s) asked. We will also learn how to **prepare a graph** comparing the variation of molarity and molality as a function of %HCl. Also, by determining the percent difference between M and m at various concentrations, we will be able to judge when it is reasonable to substitute the more convenient M for the more rigorous m in equilibrium expressions.

### Example 1.1

In the *Handbook of Chemistry and Physics* (CRC Press), percent composition and specific gravity data for solutions of many common reagent solutions are given. Let us take, for example, pairs of values for aqueous hydrochloric acid: (%HCl, sp gr at 25°C) from the Handbook and copy the data into Quattro, and then calculate the molarity, molality and mole fraction of HCl for each of these solutions.

To begin with, starting at cell **A1** and continuing in this row in successive columns, write centered labels by entering ^ before typing the label. Thus in **A1**, "Wt%", in **B1**, "Sp Gr.", in **C1**, "M" (for molarity), in **D1**, "m" (for molality), in **E1**, "X" (for mole fraction), in **F1**, "%Dif." (See Table

---

[1] Spreadsheet commands used in this book will be cited in triplicate: first, in QPRO; then, in Brackets, in QUATTRO and 123.

1.1.) Now starting with **A2,B2** enter the pair of data for each solution as given above, reaching **A18,B18** with the last pair. Place the cursor in **C2** (either by moving the cursor or, more conveniently, by using **F5**), and write the formula for molarity (remember to start with +, required for most all mathematical expressions in Q) : $+(A2/100)*B2*1000/(1.0078+35.457)$, then press Enter, to find 0.171981 in **C2**. In cells **D2..F2**, enter the appropriate formula as described above.

### Table 1.1 Partial Spreadsheet for HCl

| Wt  % | Sp Gr | M | m | X | %Dif |
|---|---|---|---|---|---|
| 1 | 1.0032 | 0.2751 | 0.2770 | 0.00497 | 0.68 |
| 2 | 1.0082 | | | | |
| 4 | 1.0181 | | | | |
| 6 | 1.0279 | | | | |
| 8 | 1.0376 | | | | |
| 10 | 1.0474 | | | | |
| 12 | 1.0574 | | | | |
| 14 | 1.0675 | | | | |
| 16 | 1.0776 | | | | |
| 18 | 1.0876 | | | | |
| 20 | 1.098 | | | | |
| 22 | 1.1083 | | | | |
| 24 | 1.1187 | | | | |
| 26 | 1.129 | | | | |
| 28 | 1.1392 | | | | |
| 30 | 1.1493 | | | | |

At this point, although we have dealt with all of the necessary formulas and data, the spreadsheet seems almost empty. This can be readily corrected by means of the Copy function. The copy command is /EC{/BC; /C}). The prompt will ask for the source block of cells. Respond by typing **B2..F2** (where you entered the formulas earlier), then Enter. You will now be asked for the destination. Enter the uppermost cell on the left (**B2**) and the lowermost right one (**F15**), thus B2..F15, then Enter. You will be rewarded by seeing the entire spreadsheet generated quickly (Table 1.2).

### Table 1.2  Aqueous Hydrochloric Acid Solution

| Wt % | Sp Gr | M | m | X | %Dif |
|------|-------|------|---------|---------|-------|
| 1 | 1.0032 | 0.2751 | 0.2770 | 0.00497 | 0.68 |
| 2 | 1.0082 | 0.5530 | 0.5597 | 0.00998 | 1.20 |
| 4 | 1.0181 | 1.117 | 1.1427 | 0.02018 | 2.26 |
| 6 | 1.0279 | 1.691 | 1.7504 | 0.03058 | 3.38 |
| 8 | 1.0376 | 2.276 | 2.3847 | 0.04120 | 4.54 |
| 10 | 1.0474 | 2.872 | 3.0471 | 0.05205 | 5.73 |
| 12 | 1.0574 | 3.480 | 3.7396 | 0.06313 | 6.95 |
| 14 | 1.0675 | 4.098 | 4.4643 | 0.07446 | 8.20 |
| 16 | 1.0776 | 4.728 | 5.2235 | 0.08603 | 9.48 |
| 18 | 1.0876 | 5.369 | 6.0198 | 0.09786 | 10.82 |
| 20 | 1.098 | 6.022 | 6.8559 | 0.1100 | 12.16 |
| 22 | 1.1083 | 6.687 | 7.7349 | 0.1223 | 13.55 |
| 24 | 1.1187 | 7.363 | 8.6601 | 0.1350 | 14.98 |
| 26 | 1.129 | 8.050 | 9.6353 | 0.1479 | 16.45 |
| 28 | 1.1392 | 8.748 | 10.6647 | 0.1612 | 17.98 |
| 30 | 1.1493 | 9.455 | 11.7530 | 0.1748 | 19.55 |

It will be useful to have a graphical display of the way in which M, m, and their %dif. varies with concentration (expressed as wt% NaCl). Review the /G pulldown menu and the section in the Manual. Use the command

/GGX{/GGX; /GG} to bring you to the graph menu and to select an XY type graph. Now select **X**, which allows you to specify the block of cells to use as X values. Respond with **A2..A15**. Press **S** (unnecessary in 123), for series (of graphs). Press **1**, or first series {A series in 123}, and select **C2..C15**; this will plot M vs. wt%. Press **2** and **3** (or B and C in 123) (for **D2..D15** and **F2..F15**) for the plots of m and %dif. (You can have as many as six plots in one spreadsheet.) Press escape to return to the graph menu, then press **T** to title the graph (you have two lines for this) as well as the X and Y axes. (See Figure 1.1) From this spreadsheet and graph, the question of how serious an error is made when molarity is used as an approximation of the molality is readily answered. Also, with the use of multiple regression (explained below), an equation suitable for calculating M or m from the wt% for all solutions of Hcl to within a desired accuracy (up to the experimental error of the Handbook data) can be obtained.

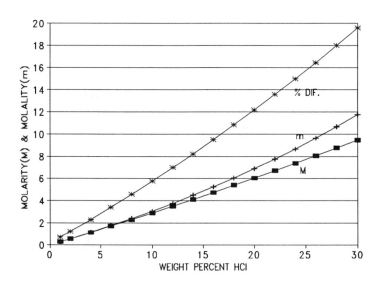

**Figure 1.1 Comparison of HCl MOLARITY and MOLALITY**

### Example 1.2 Atomic Mass Table

In this exercise, you will learn a spreadsheet technique which will permit a spreadsheet to serve as an "on line" resource that can provide a convenient substitute for a reference book. We will introduce the use of Block Names to obtain atomic and molecular weights quickly and conveniently once they are tabulated as needed. We can construct the table with random entries because the Sort function will order them according to atomic number.

First, a minor deviation from our previous practice. Instead of using ROW 1 for labels, let us start right out in this row, using COLUMN A for the symbol of the element, B for the atomic number, and C for atomic masses. The advantage of this is that the row number will now coincide with the atomic number, and you could use this to locate an atomic mass. Enter the symbol, atomic number and mass, resp., of the first thirty or so elements (here we have tabulated the first 36; see Table 1.3)

### Table 1.3 Atomic Mass Spreadsheet

| H | 1 | 1.0079 | K | 19 | 39.098 |
|---|---|--------|---|----|--------|
| He | 2 | 4.0026 | Ca | 20 | 40.08 |
| Li | 3 | 6.941 | Sc | 21 | 44.9559 |
| Be | 4 | 9.01218 | Ti | 22 | 47.9 |
| B | 5 | 10.81 | V | 23 | 50.9414 |
| C | 6 | 12.011 | Cr | 24 | 51.996 |
| N | 7 | 14.0067 | Mn | 25 | 54.938 |
| O | 8 | 15.9994 | Fe | 26 | 55.847 |
| F | 9 | 18.9984 | Co | 27 | 58.9332 |
| Ne | 10 | 20.179 | Ni | 28 | 58.7 |
| Na | 11 | 22.9898 | Cu | 29 | 63.546 |
| Mg | 12 | 24.305 | Zn | 30 | 65.38 |
| Al | 13 | 26.9815 | Ga | 31 | 69.72 |

**Table 1.3 Atomic Mass Spreadsheet (Cont'd)**

| Si | 14 | 28.086 | Ge | 32 | 72.59 |
|----|----|--------|----|----|-------|
| P  | 15 | 30.9738 | As | 33 | 74.9216 |
| S  | 16 | 32.06 | Se | 34 | 78.96 |
| Cl | 17 | 35.453 | Br | 35 | 79.904 |
| Ar | 18 | 39.948 | Kr | 36 | 83.8 |

When you are finished with a batch and you wish to **Sort**, select /DS{/ADS; /S} and respond to the prompt by listing the upper right and lower left cell addresses defining the block to be sorted. For the first key, select column **B** to sort atomic numbers numerically using the ascending mode (at another time, you may wish to designate **A** column to sort symbols alphabetically). Continue this in batches small enough to prevent boredom. Eventually, you will accumulate the data for all the elements. It might be a good idea to share the work with classmates, dividing the work and creating the complete file by team effort.

From the way in which you designed the spreadsheet, the atomic masses are listed in column **C** in order of increasing atomic number. Thus, **C1** contains the atomic mass of H, **C6**, that of C, and so forth. If you start to remember the atomic numbers of most of the elements, say, Al is #13, then obviously you recognize that its atomic mass value is in cell **C13**. The calculation of the molecular mass of NaCl can be obtained from the formula: **+C11 + C17**.

While it is useful to know the atomic numbers of some of the more commonly encountered elements, no one expects to comfortably carry all of these in his head. The spreadsheet structure makes this unnecessary. Let us use the feature known as **Block Names**. By this means, we can devise some useful and easily recognized symbol to describe a block of one or more cells that, when written, will represent the contents of that cell(s). We have a natural candidate for the suitable symbol in this case - the symbol for the element in question!

Call this feature by the command **/ENC{/BNC; /RNC}** and, at the prompt, write H. The 'block' is **C1**. Repeat **/ENC**, now entering the name He, and the 'block' **C2**. Repeat this process for all of the elements. Test the value of this exercise by entering, for a convenient and empty cell, the formula for the molecular mass of $Na_2CO_3$, **+2*$Na + $C +3*$O**. Notice that a block name is preceded by the $ sign. When you enter the formula you should find that you have the correct molecular mass formula. The usefulness of this spreadsheet technique is limited only by the student's ingenuity. After the atomic masses, further listings of GMWs of commonly encountered compounds, or even of parts of GMWs such as for $NO_3$ ($N + 3*$O), $SO_4$ ($S + 4*$O), $NH_4$ ($N +4*$H), etc. can be included. Such a spreadsheet would make calculations of gravimetric factors very simple and convenient. For example the factor for Mg in magnesium pyrophosphate is $+2^*$Mg $+2^*$P $+7^*$O).

In subsequent chapters, we will find it useful to apply the block name feature for complex equations containing a large number of variables. Using cell addresses (or absolute cell addresses for 'constants') are either inconvenient or difficult to remember throughout a lengthy calculation. Naming the blocks containing the parameters with easily remembered, logical symbols greatly simplifies such calculations (see especially Chap. 17).

### Example 1.3 Equilibrium Constant Tables

Although values for useful equilibrium constants are tabulated in the appendixes, the student will find preparing spreadsheet versions of these tables to be quite convenient.

## SOLUTION OF COMPLEX EQUATIONS

Quite frequently in chemical calculations, one encounters some rather complicated algebraic equations. While the solution to a second order (quadratic) equation has a relatively simple, general solution:

For the equation: $ax^2 + bx + c = 0$ ; $\quad x = \dfrac{-b \pm \sqrt{(b^2 - 4ac)}}{2a}$ $\qquad$ (1-6)

higher order equations are more difficult to solve. Using the spreadsheet, however, greatly simplifies the process.

## METHOD OF SUCCESSIVE APPROXIMATIONS

A time-honored method of solving complex equations is to select an approximate answer (more or less arbitrarily) to the equation and use the resulting calculation to obtain a better approximation, and repeating this until further change is within the desired limit. The tediousness and time required is reduced almost to zero with the spreadsheet, as will now be illustrated.

Consider, for example, an equation that describes the pH of a solution of 0.01 M $Na_2CO_3$: derived by methods outlined in Chapter 4 and here configured specifically for this example:

$$[H^+] = \sqrt{\frac{K_1K_w([H^+] + K_2)}{[H^+]^2 + [H^+](2C_A + K_1) + K_1(C_A + K_2) - K_w}}$$

In this form, we can see that the "solution" to $[H^+]$ is itself a function of $[H^+]$, $F([H^+])$. The method of successive approximations involves using two columns in the spreadsheet, **A** for x values, **B** for values of $F([H^+])$. Place any value of [H] in **A2**, then use the formula in the equation above in **B2**. In **A3**, use the contents of cell B2, i.e., +**B2**. Using **B2** as source, copy into **B3**; now, using **A3..B3** as source, copy to **A4..B10**. This table demonstrates how few the number of successive approximations are needed to solve this equation. Notice that when the initial guess is either 0.001 or $10^{-7}$, convergence is very rapid; you need not spend much time deciding about the first approximation!

| [H] | F([H]) |
|---------|---------|
| 0.001 | 6.08e-10 |
| 6.08e-10 | 2.56e-11 |
| 2.56e-11 | 8.50e-12 |
| 8.50e-12 | 7.43e-12 |
| 7.43e-12 | 7.36e-12 |
| 7.36e-12 | 7.36e-12 |
| 1.00e-07 | 2.82e-10 |
| 2.82e-10 | 1.81e-11 |
| 1.81e-11 | 8.06e-12 |

| 8.06e-12 | 7.40e-12 |
|----------|----------|
| 7.40e-12 | 7.36e-12 |
| 7.36e-12 | 7.36e-12 |

## Use of Absolute Cell Address

An alternate way to carry out the desired calculation adds a significant amount of convenience for very little additional effort.   Here we will introduce the spreadsheet feature called **absolute address**, which designates a cell address such as, for example, **H1**, when you intend the cell contents to be used elsewhere as a constant, as **$H$1**.  In this example, write in **F1**, "**use pH** =", and in **G1**, enter your selected value of the pH (we have selected 3 below). In **H1**, enter the formula: $+10^-G1$. In **A2**, enter **$H$1**.

One change from the table above is the editing of F([H]) using the function key, **F2.**  Simply surround the formula by  [-**@LOG(** on the extreme left and [ )] on the extreme right.  Another difference is apparent when you enter "$+10^-B2$" in A3; you will observe that the entire process will be carried out.  Henceforth, you can test the effect of changing the initial estimate of the pH by simply writing it in **G1**.

Use of the absolute cell address, like the technique of block names described above, is very convenient for adapting a single spreadsheet to solve many problems and represents an effective route to asking "What if" questions.  For example, instead of putting actual values of $C_A$, $K_1$, and $K_2$ in the formula in **B2**, place values for these parameters in, say, **I1, J1,** and **K1**, respectively, and use **$I$1** for $C_A$, **$J$1**, and **$K$1** in the expression used in **B2**.  Proceed exactly as before.  Now, however, if you want to try the effect of changing the concentration, simply enter that in **I2**.  The entire spreadsheet will be modified without any further demands on you.  The same is true when you change pK values for other $Na_2B$ solutions.  When the number of adjustable parameters gets large, however, the use of absolute cell addresses is not as convenient as block names.

|          | p{F([H])} |
|----------|-----------|
| 0.001    | 9.216198  |
| 6.08e-10 | 10.59234  |

| 2.56e-11 | 11.07033 |
|----------|----------|
| 8.50e-12 | 11.12873 |
| 7.43e-12 | 11.13297 |
| 7.36e-12 | 11.13326 |

Note that the calculated pH values shown are meant to indicate the way in which the function converges and does not pay attention to the correct number of significant figures. It should also be pointed out that sometimes the method of successive approximations is not successful. Before abandoning the method, however, a reformulation of the function should be tried.

## THE POINTER FUNCTION

The use of successive approximations can be avoided altogether by means of the **POINTER** function, which will locate all real roots of any equation within a selected interval. The use of this function has the added advantage of providing a 'global overlay' on the graphical representation of the problem under consideration.

The principle of the POINTER function can be stated very simply: If a problem can be reduced to an algebraic expression of one independent variable (pH, pM, x, etc.), called $F(x)$, which has one or more **real solutions** (roots) in a given range of interest for the variable, these can be found very simply and accurately without sacrificing any rigor.

Values of $F(x)$ are obtained, with the help of the spreadsheet, for a series of values of the independent variable, x, over the entire range of interest. By definition, the roots are those values of x at which $F(x)$ is zero. One could simply graph $F(x)$ vs. x and locate the points where the curve crosses the x axis (i.e. where x = 0). With some functions, however, the approach to zero is so gradual that the exact location is difficult to find. This problem is obviated by using the logarithmic expression and, since logarithms of negative values are not possible, using the logarithm of the absolute value of $F(x)$,**@LOG(@ABS(F(x)))**. This device, called a POINTER function (you can see why in Figure 1.2), is a convenient way to recognize roots.

A more precise solution can be achieved with the POINTER function in one of two ways: (a) customize the graph by using the "zoom lens" feature of the spreadsheet, i.e., set the scale to narrower limits, and (b) for best results, add a set of x (independent variable) values to column A in smaller increments in the region of the particular root or solution of the function, and obtain the values for the other columns by employing /EC{/BC; /C} to fill the additional cells. Before viewing the graph, place these new points with the help of the **Sort** function described in Example 1.2. When the graph is viewed again, the additional definition in the solution will be clearly marked by the pointer.

This will be applied to chemical problems throughout this text, but it is being introduced here to emphasize the point that its value is not restricted to the types of analytical chemistry, or even to chemistry in general. Examples in physics, engineering, and mathematics are equally useful.

### Example 1.4 POINTER Function with F(x)

What are the real roots of F(x) in the interval $0 < x < 20$ if $F(x) = x^6 - 45.65x^5 + 794.47x^4 - 6666.67x^3 + 27979x^2 - 54737.74x + 37942.7$? The Pointer function represents a simple and effective graphical approach to solving for the real roots of any function of an independent variable, x. Label cells **A1**, **B1**, and **C1** with x, F(x), and Pointer. Use column A to write the values of x in the desired range, 0 to 20, in intervals of 0.5 with the **/EF{/BF; /DF}** command. This will occupy **a2..a41**. In cell **B2**, enter the formula corresponding to F(x), which is: $+(a2-1.5)*(a2-3.2)*(a2-5.1)*(a2-8.7)*(a2-11.1)*(a2-16.05)$. In cell **C2**, enter the formula for the Pointer function, **@LOG(@ABS(B2))**. Now call for **/EC{/BC; /C}**, Source: **B2.C2** and destination: **B3..C41**.

Proceed to the graph function, using an XY graph in which Series 1 is **C2..C41** and x series is **A2..A41** (Figure 1.2). If we were interested in better definition of the root 1.50, we could return to the spreadsheet and add to **A42..A61**, values of x from 1 to 2 in intervals of 0.05 using **/EF{/BF; /DF}**, and copying **B2..C2** to b22..c61 with **/BC**. Before returning to the graph, let us sort the tabulated entries by using the database sort function, **/DS{/DS; /DS}**. At the prompt, indicate that the block to be sorted is **A2..C61**, using column A as the basis of an ascending sort. This will put all of the tabulated values in ascending numerical order, simplifying the operations for the resulting graph. In the graph menu, **Customize** the X series, changing **Scale** from **Automatic** to **Manual**. Use the value 1 for **Low**, and 2 for **High**, and ask for an **Increment** of 0.05. **View** and notice the improved definition of the answer indicated by the Pointer.

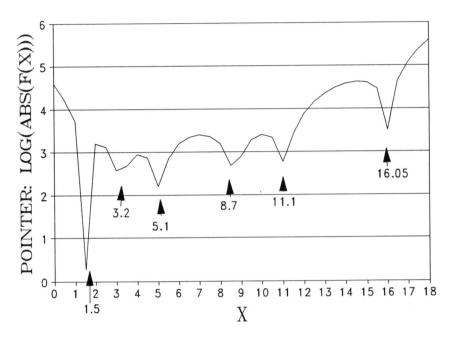

**Figure 1.2  Use of Pointer Function to Find Real Roots of F(x)**

## Representing Data by Continuous Functions: Regression Analysis

"Curve fitting", obtaining algebraic expressions which describe a set of discrete data points, is a frequently encountered, important task. For example, many analyses involve obtaining a series of measurements over a range of known concentrations of the substance being determined and fitting this data to a smooth curve which then acts as a calibration curve for determination of concentration in samples subjected to the same procedure.

In the simplest cases, the data is fitted to a linear relationship, y - mx + b.

The technique of obtaining the best fit for the data is referred to as Linear Least Squares (LS) since it represents a minimum in the sum of the squares of the deviations of each data point from that calculated from the "best" line. As recently as 20 years ago, such problems were solved without computer assistance. Today, many inexpensive hand calculators are equipped for relatively simple LS computation. Quattro provides us with a still easier method.

## Example 1.5

Consider the following set of spectrophotometric data, for copper(II), presented as concentration (in ppm), $A_{510}$ (absorbance at 510 nm) pairs: 0.0, 0.004; 1.2, 0.092; 2.0, 0.157; 3.3, 0.248; 4.5, 0.349; 5.7, 0.439; 6.7, 0.525; 8.6, 0.660; 10.1, 0.790; 12.2, 0.941; 13.9, 1.055; 16.0, 1.221; 18.2, 1.383. Enter this data in columns A and b. Now, selecting the **Advanced** item from the main menu, go to **Regression**. To summarize, key in **/TAR{/AR;/DR}**. For the **independent** variables, enter the concentrations (**A2..A14**). The **dependent** variables set are the absorbances, **B2..B14**. In the last step, select a convenient **output** location, such as **A15**. When you press **Go**, the results of the regression analysis will appear. The constant, which represents a blank reading, is 0.00559, with a standard deviation of $\pm 0.0075$; the x-coefficient, which is related to the molar absorptivity at 510 nm, is 0.07609 having a standard deviation of $\pm 0.00037$. These define the calibration line, $y = mx + b$, or

$$A_{510} = 0.07609\ C_{Cu} + 0.0059$$

When you use this equation to calculate A in column C and calculate the difference (**C2-B2**) or the % difference (100(**C2-B2**)/**B2**), a clear estimate of the reliability of the calibration can be obtained.

This simple linear regression can be readily adapted to multiple regression, i.e., to develop the best fit to an equation involving more than one variable, as in multicomponent spectrophotometric analysis (see Chapter 13), or in complex curve fitting where the results are related to the independent variable but not in a linear fashion. The well-known Taylor, McLauren, or related theorems about infinite series state that most mathematical functions, $f(x)$, can be expressed as the sum of series of terms in $x^n$. Many of these result in converging series, i.e., only several terms are necessary to represent the function with reasonable precision.

## Example 1.6

As an example, key in **/EF{/BF; /DF}** in column B with numbers from 1 to 10 (labeled as x) in steps of 0.1, and enter @LOG(B2) for all 91 cells in column A. We can now derive a series function which represents the logarithms of these numbers by entering $x^2$, $x^3$, $x^4$, $x^5$ in columns C to F (e.g., in **C2**, enter (+**B2**^2) then **/EC**). For the regression, enter **/TAR{/AR; /DR}**, and cite as independent variable **B2..F91**, as dependent

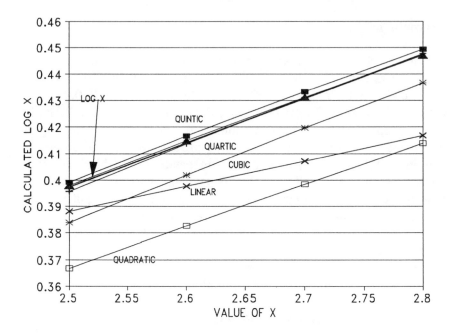

**Figure 1.3 Fitting a Function (log x) by Multiple Regression Analysis**

variable **A2..A91**. The **Output** is conveniently located at **G2**, then press **Go**. The results give not only the constant and the five "x-coefficients" by which each $x^n$ (n from 1 to 5) is multiplied, but also their standard deviations. Using the equation obtained by regression, calculate the logarithm of each of the numbers, and also the difference between the actual and calculated values. Construct an XY graph with the number as x-series, the value of log x as the first series then value of log x calculated from the equations obtained from the regression analysis using only x (linear), $x + x^2$ (quadratic), etc., all the way to the fifth order. Figure 1.3 compares all of these curve fitting expressions; the range is restricted to emphasize differences.

It will be seen that the cubic approximation gives an error of about 0.02, but the fifth order equation gives a most impressive fit. Repeat the exercise using $x^4$ as the last term, then $x^3$. How well do the simpler expressions follow the actual values?

The following equations were obtained from the regression analyses:

$$\text{LINEAR: } Y = 0.1778 + 0.0888x$$

$$\text{QUADRATIC: } Y = -0.0764 + 0.2005x - 0.00939x^2$$

$$\text{CUBIC: } Y = -0.2391 + 0.3228x - 0.0333x^2 + 0.00134x^3$$

$$\text{QUARTIC: } Y = -0.3579 + 0.4500x - 0.07468x^2 + 0.00649x^3 - 0.00022x^4$$

$$\text{5TH: } Y = -0.47469 + 0.615861x - 0.15385x^2 + 0.02304x^3 - 0.00178x^4 + 5.45E - 5x^5$$

## PROBLEMS

**1-1a**    Consult the *Handbook of Chemistry and Physics* (CRC Press) for Tables of Specific Gravities of Common Aqueous Solutions. For each of the common inorganic acids, acetic acid, and bases, as well as for an assortment of salts, develop a spreadsheet in a manner similar to that for the HCl example just completed. The class should arrange for each student to work on one or two reagents so the class can complete an extensive list of reagents to be placed in various chemistry laboratories in the department for general use.

**1-1b**    Carry out a multiple regression analysis with wt% at increasing powers as independent variables and either M or m as dependent variables. At what point, quadratic or cubic, is the M or m reproduced by the equation to within 1% relative error? Such an equation can be used to calculate M or m at concentrations not listed in the Handbook table. Does the concentration at which M and m begin to differ by 5% depend on the nature of the solute? Compare your values with those of the rest of the class.

**1-2**    Use the Pointer function approach to solving the following equations.

$$F(x) = 10x^2 - e^{x^2}.$$
Find roots from $-2$ to 2.

$$F(x) = 0.1x^5 + x^4 - 3x^3 + 2x^2 - 17x + 23$$
C Find roots from x of $-20$ to 20

1-3    Using the atomic mass spread sheets described in Example 1.2, calculate and record suitably on the spreadsheet the gravimetric factors for the following:

Fe in $Fe_2O_3$                     S in $BaSO_4$
Cl in AgCl                          Ni in $Ni(C_4H_7O_2N_2)_2$*
Pin $(NH_4)_3[P(Mo_{12}O_{40})]$    Al in Al $(C_9H_6ON)_3$**

*Ni chelate of dimethylglyoxime
**Al chelate of 8-hydroxyquinoline

1-4    The following table contains simulated spectrophotometric calibration data for two systems obtained for blanks as well as arbitrary concentrations from 1 to 10. Using regression analysis, determine whether data points for each the data from each system is better represented as a linear or quadratic relation of C.

| C | $A_{375}$ | $A_{525}$ |
|----|--------|--------|
| 0 | 0.0010 | 0.0031 |
| 1 | 0.0813 | 0.2273 |
| 2 | 0.1699 | 0.4652 |
| 3 | 0.2470 | 0.7130 |
| 4 | 0.3207 | 1.0262 |
| 5 | 0.4051 | 1.3041 |
| 6 | 0.4990 | 1.6534 |
| 7 | 0.5740 | 2.0322 |
| 8 | 0.6226 | 2.3207 |
| 9 | 0.7274 | 2.7575 |
| 10 | 0.7965 | 3.1214 |

1-5    Using the data of Table 1.2, develop an equation relating the molarity as the dependent variable and molality as the independent variable. Does a second or third order equation significantly improve the fit?

**1-6**     Begin to organize a resource spreadsheet with systematic listings
            of useful constants such as $pK_a$, $\log \beta_{MLi}$, and $E^o(pE^o)$. Restrict your
            entries to the more familiar, frequently used constants (your
            instructor will have helpful suggestions). Make use of Block names
            for convenience in locating the constant desired.

# Chapter 2

# Chemical Equilibrium

## VALUE OF EQUILIBRIUM CALCULATIONS

Chemistry, the study of chemical reactions, would be a much simpler (and less interesting) science if all chemical reactions went to completion. This is not the case, however; chemistry cannot be based on stoichiometry alone. Many reactions are reversible and do not go to completion. In such mixtures, the final composition is quite different from that which could be expected from stoichiometric considerations. A detailed knowledge of the composition, i.e., the concentration of each species present, of such mixtures is essential for the understanding of their chemistry. Short of actually measuring the amount of each and every species present, the only way to obtain this knowledge is from a consideration of the appropriate equilibria.

In planning new analytical procedures or trying to understand existing methods, questions of the following sort can be answered by information obtained through equilibrium calculations: (a) Will a substance precipitate (or separate in another manner such as by solvent extraction or ion exchange) from solution, and if so, to what extent? Will other substances separate under the same conditions? If so, can we alter conditions by using complexing agents or controlling the acidity of the solution, to obtain a selective separation? (b) What are the characteristics of reactions that are of practical use in methods of determination? Do they proceed to a sufficient extent? What are the nature and extent of any side reactions? In titrimetry, what are the criteria for the selection of suitable indicators?

In trying to interpret the physical, chemical, or biological properties of solutions, it is equally essential to know their detailed composition since each species that is present may well make a unique contribution to these properties.

The determination of equilibrium constants is naturally of utmost importance for the solution of problems of the type outlined above. In addition, the equilibrium constant provides a means of characterizing a chemical reaction and is useful in theoretical considerations. A study of equilibrium constant values provides a basis for evaluation of the influence on chemical reactions of the many factors related to chemical structure.

## REVERSIBLE REACTIONS

A reversible reaction is one that can be made to proceed in either the forward or the reverse direction by appropriately adjusting reaction conditions, such as temperature, pressure, composition, etc. In the gas phase reaction above, for example, if the pressure on the system is increased even slightly, the extent of the reaction will increase; reducing the pressure has the opposite effect.

Many oxidation-reduction reactions can be made to proceed through an external circuit to form a cell, so that an electromotive force (e.m.f.) is generated. When a greater and opposing external e.m.f. is applied across the cell, the reaction proceeds in the reverse direction. For example, the reaction

$$Cu^{2+} + Zn^{\circ} \rightleftharpoons Cu^{\circ} + Zn^{2+}$$

goes in the forward direction in the Daniell cell. Of course, if the opposing e.m.f. is exactly equal to that of the cell, then no current will flow and no chemical reaction will take place. At this point of balance only a slight change in the driving force, i.e., the opposing e.m.f., will cause a reversal of the direction of the chemical reaction. Notice that reversibility is achieved only when equilibrium position is only slightly perturbed. If the Daniell cell were in a circuit without an opposing e.m.f., significant amounts of current would flow and the system would be said to be operating under irreversible conditions.

In summary, then, a reversible reaction is one in which a slight change in driving force, such as e.m.f. (as in the Daniell cell), temperature, pressure, composition, etc. will cause a reversal of the direction of the reaction.

## FREE ENERGY AND CHEMICAL EQUILIBRIUM

The maximum useful work (in contrast to the total energy) that can be extracted from a system undergoing a reversible process is given by a thermodynamic function called the free energy, G. The work can be of several different types. A reversible expansion of a gas is a function of the pressure and the change in volume; an increase in surface area against a given surface tension; the product of an electrical potential and charge (see Chapter 7); the product of the change in free energy of a chemical reaction and the number of moles of substance undergoing the reaction.

To explain the driving force of a chemical reaction at constant temperature and pressure we must be able to describe the free energy functions of substances in various states have the following forms:

For one mole of an ideal substance in the gas phase:

$$G = G° + RT \ln f \qquad (2\text{-}1)$$

where f, the fugacity, is a dimensionless quantity used to describe gases, and is the corrected ratio of the partial pressure of the gas in the system to the partial pressure (in atmospheres) in its standard state, also unity (hypothetical unit atmospheres).

For one mole of a substance in a condensed phase, i.e., liquid or solid:

$$G = G° + RT \ln x, \qquad (2\text{-}2)$$

where x is the ratio of the mole fraction of the substance to that in its standard state, which for condensed phases is unity.

For one mole of a substance in solution:

$$G = G° + RT \ln a \qquad (2\text{-}3)$$

where a, the activity of the solute, is a corrected ratio of the concentration in the solution under study to the concentration of the solute in its standard state which is unity (hypothetical one molal), R is the universal gas constant, and T the absolute temperature. As may be seen in Equations 2-1, 2-2, and 2-3, the value of G° depends not only on the nature of the substance, but also on the state: gas, liquid, solid or solution.

Understanding the choice of standard states in a problem is critical to proper treatment. Sometimes the standard state is one which does not exist at all, but can be readily pictured, hypothetically. For example, most gas mixtures do not behave in an ideal fashion. The molecules occupy space (they are not point molecules); they will interact to some extent unless they are infinitely far apart. Hence, the commonly used standard state for gaseous substances is defined as **hypothetical** partial pressure of one atmosphere. Hypothetical, that is, because at one atmosphere, real gases will require some correction in their free energy value to compensate for their volumes and interactions. Analogously, the standard state for solutes commonly used is hypothetical one molal concentration; i.e., the concentration of an ideal solute in an ideal solution that would result in the value of the standard free energy. In real solutions, a correction would have to be applied. Such corrections are described in the next chapter.

If the change in free energy ($G_{products} - G_{reactants}$) symbolized by $\Delta G$, for a process or reaction is negative, the process or reaction will tend to proceed spontaneously (not necessarily rapidly). A positive $\Delta G$, on the other hand, corresponds to a tendency for the process to reverse spontaneously. A system is in equilibrium when $\Delta G$ is zero.

In summary then, for chemical reactions at constant temperature and pressure:

$\Delta G = -$ reaction tends to proceed spontaneously
$\Delta G = 0$ reaction is at equilibrium
$\Delta G = +$ reaction tends to reverse itself.

## FREE ENERGY AND THE EQUILIBRIUM CONSTANT

By applying the free energy function (Equation 2-3) to the general reaction system in solution:

$$aA + bB + .. \rightleftharpoons cC + dD + .. \qquad (2\text{-}4)$$

an expression which will relate the concentration variables in the equilibrium system can be derived. This expression, called the equilibrium constant expression, provides the basis for all calculations and predictions concerning the effects of changes in composition upon systems in chemical equilibrium.

For the reaction 2-4,

$$\Delta G = (cG_C + dG_D + ..) - (aG_A + bG_B + ..) \qquad (2\text{-}5)$$

In Equation 2-5, each of the free energy terms (which represent free energies per mole) has been multiplied by the appropriate number of moles in the balanced chemical equation. Substituting from Equation 2-3,

Equation 2-5 becomes:

$$\Delta G = (cG_C^\circ + cRT \ln a_C + dG_D^\circ + dRT \ln a_D + ..)$$
$$- (aG_A^\circ + aRT \ln a_A + bG_B^\circ + bRT \ln a_B +.) \qquad (2\text{-}6)$$

where the a values correspond to the activities of the components in the reaction mixture. Separating the terms which are not concentration dependent from the others, and calling these $\Delta G^\circ$, i.e.,

$$\Delta G^\circ = cG_C^\circ + dG_D^\circ + .. - aG_A^\circ - bG_B^\circ$$

we have:

$$\Delta G = \Delta G^\circ + RT \ln \{a_C^c \, a_D^d\}/\{a_A^a \, a_B^b\} \qquad (2\text{-}7)$$

At equilibrium, $\Delta G = 0$. Hence,

$$-\Delta G^\circ = RT \ln \{a_C^c \, a_D^d\}/\{a_A^a \, a_B^b\} \qquad (2\text{-}8)$$

Taking the antilogarithm of Equation 2-8, we obtain:

$$e^{-\Delta Go/RT} = {}^*K = \{a_C^c \, a_D^d\}/\{a_A^a \, a_B^b\} \qquad (2\text{-}9)$$

The exponential expression on the left-hand side of 2-9, comprised of $\Delta G^\circ$, R the gas constant, and T the absolute temperature, none of which vary with concentration, therefore is constant at constant temperature and is called the thermodynamic equilibrium constant, ${}^*K$.

Handbooks of chemistry and other appropriate reference works have extensive tabulations of $G^\circ$ values for various substances from which $\Delta G^\circ$, and therefore ${}^*K$, values for chemical reactions can be calculated. The values of $G^\circ$ of elements in their standard states (g, l, or s) at 25°C is exactly zero, as is their enthalpy $H^\circ$ (see below under Effect of Temperature).

Another useful application of Equation 2-7 is for the prediction of the direction of reaction in a mixture of any initial composition. The activity quotient (the argument of the logarithmic term) in Equation 2-7 of the mixture is calculated on the basis of the initial composition and $\Delta G$ evaluated. The direction of the reaction is now readily obtained using the sign of the $\Delta G$ as criterion. A simpler corollary of this involves comparing the value of $^*K$ for the reaction with the right-hand side of 2-8, using initial values of the activities of all the substances involved in the reaction, which is called the activity quotient.

This leads to an equivalent set of criteria for prediction of the direction of reaction based on the sign of $\Delta G$:

When $^*K >$ activity quotient the reaction tends to proceed spontaneously (composition changes so as to increase activity quotient)

When $^*K =$ activity quotient the reaction is at equilibrium (composition does not change with time)

When $^*K <$ activity quotient the reaction tends to reverse itself (composition changes so as to decrease activity quotient).

An illustration of the use of these criteria is in the development of the rules for precipitation (Chapter 6).

## RELATION BETWEEN ACTIVITY & CONCENTRATION

Although a full discussion of the concept of activity is given in Chapter 3, some remarks about the relation between activity and concentration are relevant now. In general, the activity of a substance in solution may be related to its concentration. Thus,

$$a_A = [A]\gamma_A \qquad (2\text{-}10)$$

where $\gamma_A$ is a factor which varies with the total composition of the solution and is called the activity coefficient of A. From 2-10, it follows that $^*K$ may be written as the product of K, the concentration constant, or equilibrium quotient, and the activity coefficient quotient. Thus

$$K = \frac{[C]^c[D]^d}{[A]^a[B]^b} \cdot \frac{\gamma_C^c \gamma_D^d}{\gamma_A^a \gamma_B^b} \qquad (2\text{-}11)$$

In extremely dilute solutions of most solutes, the value of each $\gamma$, and therefore, the activity coefficient quotient, approaches unity and $^*K = K$. Hence, for solutes in very dilute solutions, we may write Equation 2-9 as follows:

$$K - \frac{[C]^c[D]^d}{[A]^a[B]^b} \tag{2-12}$$

In applying equilibrium calculations to problems of chemical composition, that is, what species are present and in what concentrations, it is vital to be clear about the respective roles of the concentration of a species, [A], and its activity, $a_A$. For example, let us take 0.01 M HCl. In this solution of a strong electrolyte, the value of $[H^+]$ is unequivocally 0.01 M, and regardless of changes in the ionic strength (See Chapter 3) of this solution, obtained by adding different amounts of solid NaCl, this concentration will not change. On the other hand, the activity, $a_{H+}$, will be affected by the change in ionic strength. This points out the need to be able, in equilibrium calculations, both to properly assess the role of ionic strength on activities on all species and also to keep in mind that concentrations, not activities, are the way compositions are described. When we are measuring thermodynamic properties, as in the potentiometric measurement of pH, however, the activity must be used.

## EQUILIBRIUM EXPRESSIONS FOR VARIOUS TYPES OF REACTIONS

In developing the equilibrium expression (2-10) a reaction (2-4) was described in which all of the components were in solution. From the nature of the free energy functions given in Equations 2-1 and 2-2 it follows that any gaseous component will be represented in the equilibrium expression by its partial pressure, and any pure liquid or solid by unity, since the logarithmic term is absent in Equation 2-2. Whenever the solvent appears in the chemical equation, its free energy is considered to be sufficiently close to that of the pure liquid, provided the solutions are reasonably dilute so that it too is represented in the equilibrium expression by unity based on using mole fraction as the measure of solvent water.

The following examples will illustrate the types of equilibrium expressions usually encountered. Inasmuch as the concentration constant, K, is needed in problems involving the description of the composition, the equilibrium expressions will be described in terms of K. In all such calculations it is essential, of course, to employ the appropriate value of K, which is dependent on the ionic strength of the solution as detailed further in Chapter 3.

(a) The Dissociation of Water.

$$2H_2O \rightleftharpoons H_3O^+ + OH^-$$

$$K_w = [H_3O^+][OH^-]$$

(b) The Dissociation of Ammonia.

$$NH_3 + H_2O \rightleftharpoons NH_4^+ + OH^-$$

$$K = \frac{[NH_4^+][OH^-]}{[NH_3]}$$

(c) The Stepwise Acid Dissociation of Carbonic Acid.

$$H_2CO_3 + H_2O \rightleftharpoons H_3O^+ + HCO_3^-$$
$$HCO_3^- + H_2O \rightleftharpoons H_3O^+ + CO_3^{2-}$$

$$K_{a1} = \frac{[H_3O^+][HCO_3^-]}{[H_2CO_3]} \quad ; \quad K_{a2} = \frac{[H_3O^+][CO_3^{2-}]}{[HCO_3^-]}$$

(d)  The Stepwise Formation of the Diamminesilver (I) Complex.

$$Ag^+ + NH_3 \rightleftharpoons Ag(NH_3)^+$$
$$Ag(NH_3)^+ + NH_3 \rightleftharpoons Ag(NH_3)_2^+$$

$$K_{f_1} = \frac{[Ag(NH_3)^+]}{[Ag^+][NH_3]} \quad ; \quad K_{f_2} = \frac{[Ag(NH_3)_2^+]}{[Ag(NH_3)^+][NH_3]}$$

Note that **formation** constants, rather than **dissociation** constants, which are the reciprocals of the formation constants, are used for complex ion equilibria in accord with the practice of most workers in this field.

(e)  The Solubility of Silver Chromate.

$$Ag_2CrO_4(\text{solid}) \rightleftharpoons 2Ag^+ + CrO_4^{2-} \quad K_{sp} = [Ag^+]^2[CrO_4^{2-}]$$

(f)  An Oxidation-Reduction Reaction:

$$Fe^{2+} + Ce^{4+} \rightleftharpoons Fe^{3+} + Ce^{3+}$$

$$K = \frac{[Fe^{3+}][Ce^{3+}]}{[Fe^{2+}][Ce^{4+}]}$$

# FACTORS AFFECTING EQUILIBRIUM CONSTANTS

## Effect of Composition

At a given temperature, the value of the true, or thermodynamic, equilibrium constant, $^*K$, is independent of composition. Except in the special case of extremely dilute solutions, values of the concentration equilibrium constant, $K$, as defined by Equation 2-12 are **not** truly constant. It will vary with changes in activity coefficients, which in turn will vary with the composition of the solution. Since in most cases the activity coefficients can be calculated, $K$ values that apply to particular compositions can be calculated and are extremely useful for accurate calculations. In electrolyte solutions, the most important composition parameter determining the value of $K$ is the ionic strength (Chapter 3).

## Effect of Temperature

Suppose we are dealing with a reaction which is endothermic, i.e., one that takes place with the absorption of heat energy. A change in the conditions by adding heat energy through a temperature rise would result in an increase in the extent of this reaction inasmuch as this increase tends to minimize the effect of the temperature change. Among the examples of endothermic reactions of interest is the self-ionization of water. For example:

$$H_2O \rightleftharpoons H^+ + OH^-, \quad \Delta H^\circ = 13.8 \text{ kcal/mole}$$

where $\Delta H^\circ$ is the standard heat (enthalpy) of reaction; as might be predicted, the dissociation of water is higher at 100°C than at 20°C. Conversely, when the reaction is exothermic, the extent of reaction will decrease as the temperature increases. A number of weak acids have small heats of dissociation, e.g., $\Delta H^\circ$ for acetic acid is -0.1 kcal/mole, for formic acid -0.01 kcal/mole, for boric acid is +3.4 kcal/mole. In such instances, the extent of reaction varies only slightly with temperature. A quantitative expression of the effect of temperature upon the equilibrium constant is:

$$^*K = A\ e^{-\Delta H^\circ/RT} \qquad\qquad (2\text{-}13)$$

where the change in $^*K$, the equilibrium constant, with T, the absolute temperature, is seen to be a function of $\Delta H^\circ$. The factor A is reasonably constant over a small temperature range, and is related to the entropy change of the reaction.

## Effect of Solvents

In the majority of reactions that we will be concerned with, water is the solvent that is used. Water is distinctive in having a very high dielectric constant of 78.5 at 25°C compared with 24.3 for ethyl alcohol and 4.2 for diethyl ether. The higher the dielectric constant of a medium, the easier it is for ions to be separated in the medium, and therefore dissociation can occur more easily in water than in common organic liquids. This effect is illustrated by the manner in which the dissociation constant of acetic acid changes in a series of solvents having different dielectric constants (Table 2.1). In the same connection, note the effect of adding increasing amounts of dioxane to water on the dissociation constant of water, Table 2.2. These two examples are typical of the effect of dielectric constant upon reactions which give rise to electrical charge separation. Naturally, the reverse type of reaction, one in which charge neutralization occurs will be favored by a reduction in the dielectric constant.

All ions in solution are solvated, some to a greater extent than others. Therefore, by changing the solvent, the environment of the ions will change, in a manner that depends on the specific properties of the ions and the solvent molecules. These will determine to a large extent the size and type of solvation shell that will be formed around the ion. For example, if mixed solvents are used, such as a mixture of dioxane and water, the specific properties of the electrolyte ions may very well exert a "sorting" effect on the solvent molecules. This would mean that the electrolyte ions may be surrounded by more solvent molecules of water than dioxane. These considerations point to the fact that it may be easier for ions to exist in one type of solvent rather than in another. Therefore, the equilibrium constant for the dissociation of a substance will be dependent on solvation effects as well as on solvent 'polarity', as measured by its dielectric constant.

### Table 2.1   Effect of Solvent on the Dissociation Constant of Acetic Acid at 25°C

| Solvent | Dielectric Constant | Dissociation Constant |
|---|---|---|
| Water | 78.5 | $2 \times 10^{-5}$ |
| Methyl alcohol | 32.6 | $5 \times 10^{-10}$ |
| Ethyl alcohol | 24.3 | $5 \times 10^{-11}$ |

### Table 2.2 Ion Product of $H_2O$ in Dioxane-$H_2O$ Mixtures at 25°C

| Weight % of Dioxane | Dielectric Constant | Ion Product Constant |
|---|---|---|
| 0 | 78.5 | $1.01 \times 10^{-14}$ |
| 20 | 60.8 | $2.40 \times 10^{-15}$ |
| 45 | 38.5 | $1.81 \times 10^{-16}$ |
| 70 | 17.7 | $1.40 \times 10^{-18}$ |

### Table 2.3  Values of Thermodynamic Parameters at 25°C

| Substance (aqueous or solid) | Enthalpy H°, Kcal/mole | Free Energy $\Delta G°$, Kcal/mole |
|---|---|---|
| $NH_3$ | -19.32 | -6.37 |
| $NH_4^+$ | -31.74 | -19.00 |
| $Ba^{2+}$ | -128.67 | -134.00 |
| $BaCO_3(s)$ | -291.3 | -272.2 |
| $BaSO_4(s)$ | -350.2 | -323.4 |
| $HBr$ | -28.9 | -24.57 |
| $Ca^{2+}$ | -129.77 | -132.18 |
| $CaCO_3(s)$ | -288.45 | -269.78 |
| $CaF_2(s)$ | -290.3 | -277.7 |
| $CO_2$ | -98.69 | -92.31 |
| $Cl^-$ | -40.02 | -31.35 |
| $HCl$ | -40.02 | -31.35 |
| $Cu^{2+}$ | 15.39 | 15.53 |
| $CuS$ | -11.6 | -11.7 |
| $F^-$ | -78.68 | 66.08 |
| $H_2O$ | -70.41 | -58.21 |
| $H^+$ | 0.00 | 0.00 |
| $I^-$ | -13.37 | -12.35 |
| $HI$ | -13.37 | -12.35 |
| $Fe^{2+}$ | -21.0 | -20.30 |
| $Fe^{3+}$ | -11.4 | -2.52 |
| $Pb^{2+}$ | 0.39 | -5.81 |

| Substance (aqueous or solid) | Enthalpy H°, Kcal/mole | Free Energy G°, Kcal/mole |
|---|---|---|
| PbS(s) | -22.54 | -22.15 |
| HNO$_3$ | -49.37 | -26.41 |
| Ag$^+$ | 23.51 | 18.43 |
| AgBr(s) | -23.78 | -22.39 |
| AgCl(s) | -30.36 | -26.32 |
| AgI(s) | -14.91 | -15.85 |
| H$_2$S | -9.4 | -6.54 |
| S$^{2-}$ | 10 | 20 |
| HS$^-$ | -4.22 | 3.01 |
| HSO$_4^-$ | -211.70 | -179.94 |
| SO$_4^{2-}$ | -216.90 | -177.34 |

## PROBLEMS

2-1     Using Table 2.3, calculate $\Delta H°$ and $\Delta G°$ at 25° as well as $\Delta G°$ at 100°C. Find ˚K at 25° and 100°
        a) $Ag^+ + Cl^- \rightleftharpoons AgCl(s)$
        b) $Ag^+ + Br^- \rightleftharpoons AgBr(s)$
        c) $Ag^+ + I^- \rightleftharpoons AgI(s)$
        d) What do these data suggest as the primary reason for the decreasing solubility of the silver halides with increasing anion atomic mass?
        e) $HSO_4^- \rightleftharpoons H^+ + SO_4^{2-}$
        f) $H_2S \rightleftharpoons H^+ + HS^-$
        g) $HS^- \rightleftharpoons H^+ + S^{2-}$

2-2     Calculate H° and G° values for the OH$^-$ ion from the $\Delta H°$ and $\Delta G°$ for the self-ionization of water.

2-3      Using the fundamental thermodynamic equation that

$$G° = H° - TS°$$

where T is the absolute temperature and S°, the standard
entropy, a quantity which represents the degree of disorder or
randomness of a mole of substance in its standard state,
calculate the S° values of the substances listed in Table 2.3.

2-4      If a reaction equilibrium constant doubles when the temperature is
(a) raised 10°C (b) lowered 10°C.  What is the value of the
standard enthalpy change in each case?

2-5      From Table 2.2, graph the free energy of dissociation of water vs.
dielectric constant.  Does Coulomb's Law explain the relationship?

# Chapter 3

# The Role of Activity in Equilibrium Calculations

## ACTIVITY AND THE EQUILIBRIUM EXPRESSIONS

In Chapter 2, the relationship defining equilibrium was defined in terms of concentrations and activity coefficients:

$$* K = \frac{[C]^c [D]^d}{[A]^a [B]^b} \cdot \frac{\gamma_C^c \gamma_D^d}{\gamma_A^a \gamma_B^b} \qquad (2\text{-}11) \rightleftharpoons (3\text{-}1)$$

By rewriting Equation 3-1 as

$$K = {}^*K \cdot Q_\gamma \qquad (3\text{-}2)$$

$$\text{or} \quad K = {}^*K / Q_\gamma \qquad (3\text{-}3)$$

where $Q_\gamma$ is the activity coefficient quotient part of the right hand side of 3-1, we get Equations 3-2, 3-3 that not only show that the variation of K arises from changes in activity coefficient values, but how one obtains K values from the $^*K$ and expressions for $Q_\gamma$. In Equations 3-1 to 3-3, $Q_\gamma$ would be unity if the ions in solution behaved in an ideal manner, i.e., if they occupied no volume, had no effect on each other and were not restrained in their movement in solution in any way by neighboring solvent molecules or solute ions. Since ions carry a charge, there is a coulombic force that exists between them, but in extremely dilute solutions these forces are small because of the distance between ions, and the solution approaches ideal behavior. Hence, the activity coefficient values approach unity as the solution approaches infinite dilution.

Since we must deal with real, not ideal, solutions in which interionic forces do exist, their effects have to be corrected for in some manner if the behavior of ions in solution is to be described accurately. Since interionic forces will depend on ionic charge and concentration in solution, let us introduce the **ionic strength, I,** of a solution which measures this.

## THE IONIC STRENGTH OF A SOLUTION

The combined effect of the charges on ions and their concentrations is expressed as the **ionic strength** of a solution, defined as follows:

$$I - \sum_i^n C_i z_i^2 \qquad (3\text{-}4)$$

where $C$ is the concentration in moles per liter of an ion i, and $z$ is the charge on the ion, and **n**, the number of different ions in solution.

In calculating the ionic strength of a solution, the concentration of **all** the ions that are present in the solution and not on just the ionic species that are involved in a particular equilibrium, must be taken into account. As shown in Equation 3-7, the ionic strength of a solution is the sum of the individual contributions of **each** of the electrolytes present. It can be easily shown that the contribution to the ionic strength from each component electrolyte is proportional to its concentration multiplied by a proportionality factor, **n**, related to its charge type, i.e.,

$$\mathbf{I} = \mathbf{C} \cdot \mathbf{n} \qquad (3\text{-}4a)$$

Thus, NaCl, $HNO_3$, and HOAc (acetic acid) are examples of 1:1 electrolytes because both cation and anion are singly charged; $BaCl_2$, $Na_2SO_4$, and $H_2C_2O_4$ (oxalic acid) are 2:1 and/or 1:2 electrolytes; $MgSO_4$ is a 2:2 charge type; $FeCl_3$ represents the 3:1, and so forth. All salts of a given charge type will contribute to the ionic strength by the same multiple of their concentration in the solution.

To illustrate, in a 0.1 M in KCl aqueous solution, the concentration of both $K^+$ and $Cl^+$ is 0.1 M. Since these ions both carry unit charges, the ionic strength of the solution is given by:

$$I = (0.1 \times 1 + 0.1 \times 1) = 0.1$$

Alternatively, since KCl is a 1:1 electrolyte, from Equation 3-4a, $\mathbf{n} = 1$, and $I = M = 0.1$. The ionic strength of 0.1 M $K_2SO_4$ is equal to

$$I = (0.1 \times 2 \times 1 + 0.1 \times 2) = 0.3$$

or, since $K_2SO_4$ is a 1:2 electrolyte, $n = 3$, $I = 3 M = 0.3$.

Calculating the ionic strength of a solution containing several electrolytes such as 0.100 M in NaCl, 0.030 M in $KNO_3$, and 0.050 M in $K_2SO_4$, is simply the sum of the contributions of each of these: Since the ionic strength is given by: $= n \times M$, where $n$ is a constant for each type of electrolyte, $n$ for NaCl and $KNO_3$ is 1, and for $K_2SO_4$ is 3, therefore,

Ionic strength $= 1 \times 0.100 + 1 \times 0.030 + 3 \times 0.050 = 0.280$.

You will find it helpful to work out for yourself what this proportionality constant, $n$, is for electrolytes of other charge types.

Care must be taken in evaluating the ionic strength contribution of weak electrolytes. For example, if any of the solutions above contained phosphoric or acetic acid also, the ionic strength would be essentially the same, because only the dissociated phosphoric or acetic acids contributes, and this is generally very small. If, on the other hand, $H_3PO_4$ is the only solute present, then an approximate equilibrium calculation must be carried out (see Example 3.3), an ionic strength calculated, and the process repeated until values of I remain constant. This may take one or two successive approximations.

# THE DEBYE-HÜCKEL THEORY OF STRONG ELECTROLYTES

It is possible to understand why solutions of electrolytes do not behave in an ideal manner in terms of both the coulombic attraction on ions which serves to constrain their movement and the thermal agitation which counteracts this restraint. Debye and Hückel developed a theory in which electrostatic forces shaping the behavior of the ions in solution as well as their finite radii formed a basis from which expressions for the activity coefficient of an ion could be derived. One of the simpler usable equations they developed, referred to as the Extended Limiting Law, gives the activity coefficient, $\gamma$, of an ion i, having a charge $z_i$ in a solution of ionic strength I.

$$-Log\, \gamma_i = \frac{Az_i^2\sqrt{I}}{1 + Ba\sqrt{I}} \qquad (3\text{-}5)$$

A and B are constants and equal to 0.51 and $3.3 \cdot 10^7$, respectively, at 25° in aqueous solution; a is the ion size parameter which is a measure of the diameter of the hydrated ion. In Equation 3-8, the value of I can be calculated for any solution, but the ion size parameter, a, has to be known before the activity coefficient of an ion can be calculated. In many calculations, the ion size parameter may be taken to be about 3Å and, therefore, $a \cdot B = 3 \cdot 10^{-8} \cdot 0.33 \cdot 10^8$, which makes $a \cdot B$ approximately equal to unity. Equation 3-5 then becomes:

$$-Log\, \gamma_i = \frac{0.51\, z_i^2\sqrt{I}}{1 + \sqrt{I}} \qquad (3\text{-}6)$$

Equations 3-5 and 3-6 apply only to dilute solutions ($< {\sim}0.02$ M). Although in extremely dilute solutions, the denominator in Equation 3-6 approaches unity, and this equation reduces to its simplest possible form, 3-7, referred to as the Debye-Hückel Limiting Law. The limiting law is truly limiting and not a very useful law, however. Some chemists say it is only useful in "slightly dirty water".

$$-\log \gamma = 0.51 \cdot z_i^2 \sqrt{I} \qquad (3\text{-}7)$$

The Extended Limiting Law, Equation 3-6, fails well before the I reaches 0.1. At values of I of 0.05 or more, a semi-empirical expression, considered by Debye and Hückel, but now known as the Davies' Equation,

$$Log\, \gamma_i = z_i^2 \left[ 0.15I - \frac{0.51 \cdot \sqrt{I}}{1 + \sqrt{I}} \right] = z_i^2 \cdot I' \qquad (3\text{-}8)$$

where $I'$ symbolizes the expression in brackets, must be used. Even the Davies' equation fails in solutions whose ionic strength is about 0.2 or higher because of the complexities in evaluating electrostatic interactions in such media.

The student will find it useful and convenient to prepare a table of values of $I'$ at ionic strength values from 0 to 1 in steps of 0.01. Perhaps preparing this spreadsheet by using three columns: A, C, and E to list values of ionic strength in rows of about thirty-three items each would make a hard copy that could be easily pasted in your notebook. If you are using QPro, then another, perhaps even more convenient, way to utilize this table in your problem solving, would be to **LINK** this QPro file to the one being used for the problem at hand. You could then call up the activity coefficient table as needed without leaving your problem spreadsheet.

## ACTIVITY COEFFICIENT CORRECTIONS

The use of activity coefficient corrections is readily accessible by means of development of spreadsheet files incorporating the Debye-Hückel limiting law (L.L.), the extended limiting law, or the Davies' equation. This is accomplished by using the /EF{/BF;/DF} command in column **A** with suitably spaced values of the ionic strength, e.g., from 0.01 in 0.01 increments to 1.00, followed by the @SQRT(A2) in column **B**, and the appropriate formulas for the L.L., extended L.L., and the Davies' equation, respectively, in the next three columns. The graphical display of the logarithms of the activity coefficients calculated according to these three equations as a function of ionic strength is shown in Fig. 3.1 as are the differences of the L.L. and the extended L.L. from the Davies' equation.

**Example 3.1**

Calculate the activity coefficient of the $Pb^{2+}$ ion in (a) a solution which is 0.005 M in $Pb(NO_3)_2$ and (b) a solution which is 0.005 M in $Pb(NO_3)_2$ and 0.040 M in $KNO_3$. Compare the values from L.L., ext. L.L., and Davies' Equation.

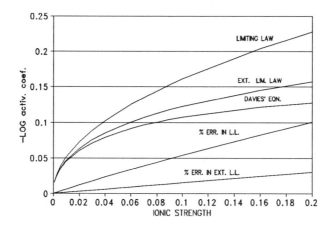

**Figure 3.1 Comparison of Activity Coefficient and Expressions**

(a) The value of $I = 3 \cdot 0.005 = 0.015$ since this is a 2:1 electrolyte. Using the Davies' equation

$$\log \gamma_{Pb} = -0.217$$

Note that the L.L. value is $-0.250$, low by 0.033 log units, and that the ext. L.L. value of $-0.223$ is low by only 0.006.

(b) The value of $I = 0.015 + 0.04 = 0.055$

The Davies' equation value is

$$\log \gamma_{Pb} = -0.339$$

Here the L.L. value, $-0.433$, is low by 0.094, and the ext. L.L. value, $-0.357$ by 0.018. Keep in mind that these are logarithmic differences; 0.094 is an error corresponding to 24.2% in the activity coefficient.

The significant influence of the charge of an ion and of the ionic strength upon activity coefficient values is obvious from Equations 3-6 to 3-8. However, there are two other factors that affect the values of activity coefficients that also merit our interest, namely temperature and nature of solvent. In order to understand these effects, let us return to Equation 3-8 and observe that the parameter A is inversely proportional to the product of the dielectric constant, D, and the absolute temperature T raised to the 3/2 power.

$$A = \frac{CONSTANT}{D^{3/2} \cdot T^{3/2}} \qquad (3\text{-}9)$$

The parameter A is not too sensitive to temperature changes. A change in temperature of 5°C from 25°C will cause a change in A of less than 1%. Changing solvents would give rise to much more serious changes in activity coefficients since the resulting change in dielectric constant is relatively large. In a 50% v/v mixture of ethanol and water, the dielectric constant has a value of 49. Hence, the parameter A in this mixture is about twice its value in water.

## ACTIVITY COEFFICIENTS OF IONS

In the preceding section, a theoretical approach to the evaluation of ionic activity coefficients was developed and was based on the Debye-Hückel theory, which was experimentally verified in dilute solutions of electrolytes. In solutions of ionic strengths greater than 0.15 to 0.2, empirical values of activity coefficients must replace those calculated from Equation 3-8 to provide an adequate accuracy.

Activity coefficients of single ions cannot be measured directly. Instead, on the basis of experiments in which the free energy of an electrolyte is determined by various methods (such as by measuring freezing point depressions of solutions or by the measurement of the e.m.f.'s of cells), a quantity called the mean ionic activity coefficient is obtained. For an electrolyte $A_m B_n$, the mean ionic activity coefficient is defined as follows:

$$(\gamma_\pm)^{m+n} = (\gamma_A)^m \cdot (\gamma_B)^n$$

where $\gamma_A$ and $\gamma_B$ are the ionic activity coefficients. However, Kielland (J. Kielland, *J. Am. Chem. Soc.*, 59, 1675, 1937) has developed a series of values of single ion activity coefficients from which values of $\gamma$ for various electrolytes may be calculated which are in reasonable agreement with experimental values up to an ionic strength of about 0.1.

## ACTIVITY COEFFICIENTS OF MOLECULAR SOLUTES

The activity coefficients of uncharged solutes are described fairly accurately up to unit ionic strength by the equation:

$$\log \gamma = k \times I$$

in which $\gamma$ and I have their usual meaning, and k is a proportionality constant called the salting coefficient (or Setchenow Constant) which depends on the nature of both the solute and the ions in solution.

**Table 3.1 Salting Coefficients for Various Neutrals**

| Salt | $k_{Acetic\ Acid}$ | $k_{Benzoic\ Acid}$ | $k_{CO2}$ |
|------|------|------|------|
| LiCl | 0.075 | | |
| NaCl | 0.066 | 0.191 | |
| KCl | 0.033 | 0.152 | 0.059 |
| KBr | 0.02 | | 0.045 |
| NaOAc | -0.014 | | |
| $KNO_3$ | -0.020 | 0.025 | 0.025 |

Since values of k vary for the most part between 0.01 and 0.10, activity coefficients of molecular solutes can be considered to be effectively unity in solutions whose ionic strengths are $\leq 0.2$.

## ACTIVITIES AND EQUILIBRIUM CALCULATIONS

As mentioned above, in solving equilibrium problems, both concentrations and activity coefficients of the substances participating in the equilibrium must be evaluated. Implicit in Equation 3-3 is the basis for a very important approach to the incorporation of activity coefficients in all equilibrium calculations. As will be discussed below, it will be possible to arrive at a sufficiently good value of the activity coefficient quotient, $Q_\gamma$, in terms of a single experimental parameter, the ionic strength. Hence, a useful strategy for taking activities into account in equilibrium calculations consists of:

1.           Obtain the value of the ionic strength of the solution. This, together with the tabulated $^*K$ value, are combined in a suitable one of the equations from 3-11 to 3-15 to arrive at the value of K which applies to the problem at hand.

2.           The remainder of the equilibrium calculation is conducted as described in subsequent chapters, but without the need of referring ever again to activity coefficients until after the concentrations have been calculated.

3.           When this is done, use of the appropriate activity coefficient readily converts the equilibrium concentrations of the species to their activities as needed.

THROUGHOUT THE REMAINDER OF THIS BOOK, EXCEPT WHERE THE DISCUSSION DEMANDS IT, WE WILL FOLLOW THE PRACTICE OF USING THE SYMBOL K, RATHER THAN $^*K$, TO REPRESENT THE APPROPRIATE VALUE. IN NUMERICAL SOLUTIONS TO THE EXAMPLES AND PROBLEMS, USE OF $^*K$ OR THE ACTIVITY-CORRECTED K, WILL BE OPTIONAL. Such corrections are essential for reasonably accurate answers, but their incorporation into the course will be left in the instructor's hands.

From Equation 3-8, which provides us with a means of calculating individual activity coefficients of ions as functions of ionic strength which can be combined to give the activity coefficient quotient, $Q_\gamma$, a single factor which is a function of the ionic strength. Thus, the value of K for any equilibrium reaction may be written as a function of ionic strength.

Considering the logarithmic form of Equation 3-3, we may write

$$\text{Log } K = \text{Log } ^*K - \text{Log } Q_\gamma \text{ or } pK = p^*K + \text{Log } Q_\gamma$$

Substituting using Equation 3-8 for values $\text{Log } Q_\gamma$ in terms of the $\gamma$,

$$pK = p^*K + (cz_C^2 + dz_D^2 - az_A^2 - bz_B^2) \cdot I' \qquad (3\text{-}10)$$

where $I'$ has been defined in 3-8, but could also be taken from the Extended limiting law (Equation 3-6). Note that if A, B, C or D is uncharged, $z = 0$ and $\log \gamma = 0$.

From Equation 3-10, the usefulness of this approach to evaluating $Q_\gamma$ is evident since $(az_A^2 + bz_B^2 - cz_C^2 - dz_D^2)$ will be a constant multiplier having a value depending on the type of equilibrium equation that is under consideration.

For example, consider the solubility equilibrium of a slightly soluble electrolyte $A_mB_n$.

$$A_mB_n \rightleftharpoons mA^{n+} + nB^{m+}$$

For this reaction, using Equation 3-10, where $c = m$, $d = n$, $z_A = n$, $z_B = m$ and $a = 0$, $b = 0$, $z = 0$ and $z = 0$, we obtain:

$$pK_{sp} = p^*K_{sp} + (mn^2 + nm^2) \cdot I' \tag{3-11}$$

For a 1:1 electrolyte such as AgCl,

$$pK_{sp} = p^*K_{sp} + 2.0\ I' \tag{3-12}$$

For a 1:2 electrolyte, such as $Ag_2CrO_4$ or $PbCl_2$,

$$pK_{sp} = p^*K_{sp} + 6 \cdot I' \tag{3-13}$$

For a 2:2 electrolyte, such as $BaSO_4$,

$$pK_{sp} = p^*K_{sp} + 8 \cdot I' \tag{3-14}$$

A similar series of expressions may be derived for an uncharged polyprotic acid, such as $H_nX$.

$pK_{a1} = p^*K_{a1} + 2I'$ (Also applicable to an uncharged monoprotic acid.)

$$pK_{a2} = p^*K_{a2} + 4I'$$
$$pK_{a3} = p^*K_{a3} + 6I'$$

For a singly charged cationic acid, such as the ammonium ion, K is equal to $^*$K at all ionic strengths.

With metal complex equilibria, where formation rather than dissociation constants are employed (Chapter 5), the expression looks a little different because $\log \beta$ and not $p\beta$ is the customary notation. For the formation of

$FeY^-$ from $Fe^{3+}$ and EDTA anion, $Y^{4-}$, for example,

$$\log \beta_{FeY} = \log{}^*\beta_{FeY} + (3^2 + 4^2 - 1^2)I' \text{ or}$$

$$\log \beta_{FeY} = \log{}^*\beta_{FeY} + 24I'$$

In general,

$$pK = p^*K + NI' \qquad (3\text{-}15)$$

where N is an integer whose value depends upon the nature of the equilibrium involved as illustrated above. As may be seen, activity corrections to equilibrium constants can be quite high especially as N and I increase.

## Example 3.2

What is the $pK_2$ of carbonic acid in a solution of ionic strength 0.050 at 25°C?

$$pK_2 = p^*K_2 + 4I'$$

From Equation 3-11, $pK_2 = pK_2 + 2 \cdot (-0.183)$. Since $p^*K_2 = 10.33$, $pK_2 = 9.96$.

It is interesting to realize that $I'$ is negative over a fairly large range of ionic strength values (certainly for $I \leq 1$) resulting in $pK < p^*K$. Thus, in all of the equations relating pK to $p^*K$, the value of pK decreases with increasing ionic strength. This will be true in all equilibria in which there is a net charge separation. In all equilibria of this type, therefore, the extent of the reaction will always increase with increasing indifferent electrolyte concentration, i.e., with the concentration of an electrolyte that does not participate in the equilibrium. Hence, weak neutral acids become stronger, slightly soluble salts become more soluble, with increasing ionic strength. This effect, although significant, is swamped whenever the electrolyte is directly involved in the equilibrium.

## Example 3.3

Calculate $K_a$ values for a 1.0 M solution of $H_3PO_4(p^*K_1=2.148)$.

While most problems in equilibrium calculations involve solutions in which the contribution of weak electrolytes to the ionic strength needed to obtain a valid K value, is negligible, a calculation of the pH of $C_A$ M HOAc or $C_A$ M $H_3PO_4$ requires finding the extent of dissociation in each case since the ionic strength value is obtained directly from these.  The extent of dissociation, in turn, cannot be calculated without having suitable, valid $K_a$ values.  The way out of this seemingly difficult impasse can be found by using successive approximations to $K_a$, as described in Chapter 1.

As will be more fully understood following Chapter 4, the only dissociation that need be considered in this $H_3PO_4$ is the first

$$H_3PO_4 \rightleftharpoons H^+ + H_2PO_4^-$$

Hence, $H_3PO_4$ is 1:1 electrolyte, and the ionic strength equals $[H^+]$. From Chapter 4, the appropriate equation for $[H^+]$ is the quadratic:

$$[H^+] = 1/2(\sqrt{Ka_1^2 + 4Ka_1 C_A} - K_{a_1}) \tag{3-16}$$

By using this equation to solve for $[H^+]$, we have, as a first approximation, the ionic strength ($[H^+] = I$). Now we can modify the pK value by means of the activity correction through several **successive approximations** until the final pH does not vary by more than $\pm$ 0.01.

Thus, in the Table below, cell A1 contains the title of the table, cells A2 to D2 carry the labels for this problem.  Cell A3 has the $^*$K value ($10^{-2.148}$), cell B3 has the formula representing the solution of the quadratic equation describing the $[H^+]$ in a relatively strong weak acid, cell C3 has the Davies' equation at an ionic strength equal to $[H^+]$, which is the concentration of dissociated acid, and D3 is the value of pH corresponding to the $[H^+]$ value in B3.  In cell A4, write the formula $10^{-2.148 + C3}$, the activity corrected value. For B4 to D4, /EC{/BC; /C} the values from B3 to D3. Now execute /EC{/BC; /C} from A4..D4 as source, to A5..D9, and the table as shown will result.  Notice how quickly the successive approximations converge.

| ACTIVITY CORRECTIONS IN 1 MH3PO4 | | | |
|---|---|---|---|
| $K_{a1}$ | [H] | I' | pH |
| 0.007112 | 0.080852 | -0.20157 | 1.092307 |
| 0.011313 | 0.100855 | -0.21560 | 0.996302 |
| 0.011684 | 0.102409 | -0.21656 | 0.989663 |
| 0.011710 | 0.102516 | -0.21662 | 0.989208 |
| 0.011712 | 0.102524 | -0.21663 | 0.989176 |

The biggest change, one that would in this case cause an error of 0.09 pH units occurs when we perform the first refinement. The second approximation changes the pH by 0.01 to 0.99 and there is no further significant change.

In this solution, whose I' = 0.2166, the values of pKa$_2$ (p*Ka$_2$ + 4I') and pKa$_3$ (p*Ka$_3$ + 6I') are 6.33 and 11.1, respectively.

## PROBLEMS

**3-1** Develop a spreadsheet showing values of activity coefficients for mono-, di- and tri-valentions using a) the Debye-Hückel Limiting Law (L.L.), b) the extended L.L., and c) the Davies' equation in solutions whose ionic strength varies from 0.001 to 0.200. Plot the above as well as the error in activity coefficient (defined as the difference of the value from that obtained with the Davies' Equation) when a) the L.L. and b) the extended L.L. is used over the specified range.

**3-2** For the following equilibrium constants develop relations between log *K and log K as a function of I' analogous to Equations 3-10 and 3-11:

(a)     $Ni^{2+} + 6NH_3 \rightleftharpoons Ni(NH_3)_6^{2+}$
(b)     $Cd^{2+} + 4Cl^- \rightleftharpoons CdCl_4^{2-}$
(c)     $Fe^{3+} + 6F^- \rightleftharpoons FeF_6^{3-}$
(d)     $Cu^{2+} + Y^{4-} \rightleftharpoons CuY^{2-}$

**3-3** Calculate the ionic strength at 1 mL intervals in V, the titrant volume, from zero to 60 mL in the following titrations:

a)     50 mL 0.01 M $Na_2CO_3$ with V mL of 0.02 M HCl.
b)     50 mL 0.1 M HOAc with V mL of 0.1 M NaOH. (At V = 0 use successive approximations (example 3-3) to find I.

**3-4** Calculate the pK values using the Davies' equation as a function of ionic strength for the titrations in problem 3-2.

3-5    Table 3.1 describes the experimentally determined activity
coefficients of a series of alkali metal halides at 25°C.

a)    Plot these data as a function of ionic strength (same as **m** for 1:1
salts)
b)    Compare the $\gamma$ values in Table 3.1 with those calculated for the
Davies' equation. Plot the error in log $\gamma$ obtained by using the
Davies' Equation as a function of ionic strength.
c)    Using multiple regression (see Chapter 11), try to fit each set of
data to a function of $I^{1/2}$, $I$, and $I^{3/2}$ as the three independent
variables. Compare the resulting equations to the comparable
Davies' equations.

**Table 3.1 Experimental $\gamma$ Values of Alkali Metal Halides at 25°C**

| m | LiCl | NaCl | KCl | LiBr | NaBr | KBr | LiI | NaI |
|---|------|------|-----|------|------|-----|-----|-----|
| 0.1 | 0.792 | 0.778 | 0.769 | 0.794 | 0.781 | 0.771 | 0.811 | 0.788 |
| 0.2 | 0.761 | 0.734 | 0.717 | 0.764 | 0.739 | 0.721 | 0.800 | 0.752 |
| 0.3 | 0.748 | 0.710 | 0.687 | 0.757 | 0.717 | 0.692 | 0.799 | 0.737 |
| 0.5 | 0.742 | 0.682 | 0.650 | 0.755 | 0.695 | 0.657 | 0.819 | 0.726 |
| 0.7 | 0.754 | 0.660 | 0.626 | 0.77 | 0.687 | 0.637 | 0.848 | 0.729 |
| 1.0 | 0.781 | 0.658 | 0.605 | 0.811 | 0.687 | 0.617 | 0.907 | 0.739 |
| 1.5 | 0.841 | 0.659 | 0.585 | 0.899 | 0.704 | 0.601 | 1.029 | 0.772 |
| 2.0 | 0.931 | 0.671 | 0.575 | 1.016 | 0.732 | 0.596 | 1.196 | 0.824 |
| 2.5 | 1.043 | 0.692 | 0.572 | 1.166 | 0.77 | 0.596 | 1.423 | 0.889 |
| 3.0 | 1.174 | 0.72 | 0.573 | 1.352 | 0.817 | 0.600 | 1.739 | 0.967 |
| 3.5 |  | 0.753 | 0.576 |  | 0.871 | 0.606 |  | 1.060 |
| 4.0 |  | 0.792 | 0.582 |  | 0.93 | 0.615 |  |  |

# Chapter 4

# Acid Base Equilibrium

---

## THEORIES OF ACIDS AND BASES

Of the many theories that have been proposed through the years to explain the properties of acids and bases, the Brønsted-Lowry, or proton transfer theory, and the older, more general Lewis theory are most generally useful.

### The Brønsted-Lowry Theory

In 1923 Brønsted and Lowry each developed an acid-base theory based on the central role of the proton. They defined an acid as a proton donor and a base as a proton acceptor. Thus, an acid-base reaction is one in which proton transfer occurs, i.e.,

$$\text{Acid} \rightleftharpoons \text{Base} + H^+ \qquad (4\text{-}1)$$

According to this definition, neutral molecules such as $H_3PO_4$ or $H_2O$, cations such as $NH_4^+$, and anions like $H_2PO_4^-$ all behave as acids, e.g.,

$$NH_4^+ \rightleftharpoons NH_3 + H^+ \qquad (4\text{-}2)$$

Similarly, cations ($H_2NCH_2CH_2NH_3^+$), anions ($HC_2O_4^-$), and neutral molecules can all act as bases. Certain substances such as $H_2O$ and $SH^-$ behave as acids as well as bases, and are called ampholytes or amphoteric electrolytes.

$$\underset{\text{acid}}{SH^-} \rightleftharpoons H^+ + \underset{\text{base}}{S^=}$$

$$H^+ + \underset{\text{base}}{SH^-} \rightleftharpoons \underset{\text{acid}}{H_2S}$$

52

Equation 4-2 is a simplification of the proton transfer reaction that takes place if the reaction is carried out in a solvent such as water. Bare protons do not exist in any solvent. The characterization of a substance as an acid or a base may be made in terms relative to the solvent water. That is to say, an acid is a substance capable of donating a proton to water, and a base is a substance capable of accepting a proton from water. Reaction 4-2 is therefore more correctly represented as follows:

$$NH_4^+ + H_2O \rightleftharpoons NH_3 + H_3O^+ \qquad (4\text{-}3)$$

The hydrated proton is represented as $H_3O^+$. Although it is, strictly speaking, incorrect to write $H^+$ to represent a hydrated proton, this is generally accepted for the sake of convenience. Throughout this book, $H^+$ and $H_3O^+$ will be used interchangeably.

It is evident from Equation 4-3 that there is a proton transfer from the acid, $NH_4^+$, to the water molecule, $H_2O$. Therefore, there are two substances that behave as acids and two substances that behave as bases in this reaction. The cation $NH_4^+$ and the hydrated proton $H_3O^+$ are both acids, while the neutral molecule $H_2O$, as well as the ammonia molecule $NH_3$, are bases.

$$NH_4^+ + H_2O \rightleftharpoons NH_3 + H_3O^+ \qquad (4\text{-}4)$$

$$acid_1 + base_2 \rightleftharpoons base_1 + acid_2$$

The pairs of compounds $NH_4^+$ - $NH_3$ and $H_3O^+$ - $H_2O$ are called **conjugate acid-base pairs**. Water can lose a proton as well as gain one and is therefore called an **amphiprotic substance**.

$$H_2O + H_2O \rightleftharpoons OH^- + H_3O^+ \qquad (4\text{-}4a)$$

$$acid_1 + base_2 \rightleftharpoons base_1 + acid_2$$

Equation 4-4a represents a type of reaction called **autoprotolysis** which, as we will see later, is important in describing the utility of solvents as acid-base titration solvents.

Further examples of acid-base reactions are:

Dissociation of weak acids:

$$CH_3COOH + H_2O \rightleftharpoons CH_3COO^- + H_3O^+ \tag{4-5}$$

$$NH_4^+ + H_2O \rightleftharpoons NH_3 + H_3O^+ \tag{4-5a}$$

Dissociation of weak bases:

$$NH_3 + H_2O \rightleftharpoons NH_4^+ + OH^- \tag{4-6}$$

$$CH_3COO^- + H_2O \rightleftharpoons CH_3COOH + OH^- \tag{4-6a}$$

Since all of these reactions involve conjugate acid-base pairs, which are completely described by $K_a$, once the $K_a$ is known, the character, $K_b$, can be derived from it as seen below.

The autoprotolysis (self dissociation) constant of water (Equation 4-4a) is given by

$$K_w = [H_3O^+] [OH^-] \tag{4-7}$$

which at 25° C and zero ionic strength is $10^{-14.00}$. The corresponding values of autoprotolysis constants for methanol, ethanol, acetic acid, and formic acids expressed as their negative logarithms (14 for $H_2O$) are 16.7, 19.1, 14.5, and 6.2, respectively. A 0.10 M "strong" acid solution in each solvent would give a concentration of the conjugate base of each solvent as $10^{-15.7}$, $10^{-18.1}$, $10^{-13.5}$, and $10^{-5.2}$.

These numbers would also represent the concentrations of the conjugate acid form of each solvent containing 0.1 M "strong" base. Therefore, water and all of the solvents except formic acid are useful acid-base titration solvents because a wide range of concentrations of conjugate acid/base species is essential. In the case of formic acid, the protonated and deprotonated solvent species do not get small enough to provide the necessary range needed for good titrations.

The equilibrium constant for Equation 4-5a is given where $K_a$ is the acid dissociation constant of the acid. The equilibrium constant for Equation 4-6a is given by:

$$K_b = \frac{[NH_4^+][OH^-]}{[NH_3]} \qquad (4\text{-}8)$$

This constant can be readily related to $K_a$ and $K_w$ so that there is no need for separate tables of $K_b$ values where $K_b$ is termed the dissociation constant of a base. Equation 4-6a may be recognized as a result of subtracting the acid dissociation reaction of $NH_4^+$, from the dissociation reaction of water. Thus:

$$H_2O \rightleftharpoons H^+ + OH^-$$
$$-(NH_4^+ \rightleftharpoons H^+ + NH_3)$$
$$\overline{NH_3 + H_2O \rightleftharpoons NH_4^+ + OH^-}$$

This subtraction of equations is equivalent to dividing the corresponding equilibrium expressions.

$$\frac{K_w}{K_a} = \frac{[H^+]\,[OH^-]}{[NH_3]\,[H^+]/[NH_4^+]} = \frac{[NH_4^+]\,[OH^-]}{[NH_3]} \qquad (4\text{-}9)$$

Hence, we see that $K_b$ from 4-8 is identical to $K_w/K_a$.

Although this relation was derived from a specific consideration of the $NH_4^+$ / $NH_3$ conjugate pair, it will be recognized as a generally valid reaction, that is,

$$K_a \cdot K_b = K_w \qquad (4\text{-}10)$$

From this equation, it follows readily that if we list a series of acids in increasing strength, we have automatically listed the conjugate bases in the order of decreasing strength.

## The Lewis Theory

The Lewis Theory of acids and bases does not feature the special role that the proton has in the Brønsted-Lowry theory. This results from defining an acid as any electron-pair deficient species. A base from this viewpoint is a species capable of furnishing electron pairs. Thus, acid-base reactions are considered as coordination reactions. This theory is of great value in understanding metal coordination complex formation.

In the following reactions:

$$H^+ + NH_3 \rightleftharpoons NH_4^+$$

$$Ag^+ + 2NH_3 \rightleftharpoons Ag(NH_3)_2^+$$

the $Ag^+$ ion* is seen to act in a manner that is similar to a proton, and the $Ag^+$ ion behaves as a dibasic acid. The subject of metal coordination complexes which involves reactions of this type will be considered at length in Chapter 5. The Lewis Theory has also found extensive use in explaining reactions such as the Friedel-Crafts reaction in nonaqueous media, involving nonprotonic acids such as $BF_3$, $AlCl_3$, etc.

## STRENGTHS OF ACIDS AND BASES

The strengths of acids, that is their proton donating tendencies, naturally depends upon the proton-accepting tendency of the base with which they react. An ordering of acids according to their strengths is obtained by the use of a reference base. In aqueous solutions, the reference base is water. Thus, the strengths of a series of acids, $HSO_4^-$, $H_2CO_3$ and $HCN$ can be compared by measuring the extent of their dissociation in water.

$$HSO_4^- + H_2O \rightleftharpoons H_3O^+ + SO_4^{2-}$$

$$H_2CO_3 + H_2O \rightleftharpoons H_3O^+ + HCO_3^-$$

$$HCN + H_2O \rightleftharpoons H_3O^+ + CN^-$$

$$(acid_1 + base_2 \rightleftharpoons acid_2 + base_1)$$

Similarly, the comparison of the strengths of bases in aqueous solutions involves the use of water as the reference acid.

$$NH_3 + H_2O \rightleftharpoons NH_4^+ + OH^-$$

$$C_6H_5NH_2 + H_2O \rightleftharpoons C_6H_5NH_3^+ + OH^-$$

aniline

$$C_5H_5N + H_2O \rightleftharpoons C_5H_5NH^+ + OH^-$$

pyridine

$$(CH_3)_3N + H_2O \rightleftharpoons (CH_3)_3NH^+ + OH^-$$

trimethylamine

In these reactions, there is a competition for protons between the two bases. Therefore, the relative amounts of the conjugate acid-base pairs that exist at equilibrium will be a measure of the strengths of the acids and bases. This is equivalent to saying that the dissociation constants measure strengths of acids and bases.

The acid dissociation constants for the acids $HSO_4^-$, $H_2CO_3$, and HCN are written as follows:

$$K_a = \frac{[H_3O^+]\,[SO_4^{2-}]}{[HSO_4^-]} = 1.0 \cdot 10^{-2}$$

$$K_a = \frac{[H_3O^+]\,[HCO_3^-]}{[H_2CO_3]} = 4.4 \cdot 10^{-7} \qquad (4\text{-}11)$$

$$K_a = \frac{[H_3O^+]\,[CN^-]}{[HCN]} = 4.0 \cdot 10^{-10}$$

It is obvious that the larger the numerical value of $K_a$, the stronger the acid, i.e., the tendency to lose a proton is greater. Therefore, the acids can be arranged in order of decreasing strength as follows:

$$HSO_4^- > H_2CO_3 > HCN$$

It is of interest to note that the values of $pK_a$ (defined as $pK_a = -\log_{10}K_a$), increase with decreasing acid strength. This means that $pK_a$ values increase with increasing strengths of the conjugate bases. For this reason, values of $pK_a$ are often used as measures of basic strength or basicity.

Unfortunately, it is not possible to establish the relative strengths of strong acids or of strong bases in this manner. Strong acids such as HCl,

$HNO_3$, and $HClO_4$, all appear equally strong when dissolved in water, because they react quantitatively with water to yield the ion $H_3O^+$ in each case. This is referred to as the leveling effect. The relative strengths of these acids can, however, be determined in solvents less basic than water in which incomplete reaction occurs. Similarly, strong bases can be differentiated in solvents less acidic than water.

## The Concept of pH

In a great many examples of practical importance, it is necessary to deal with small concentrations of hydrogen or hydroxyl ions; the method of writing these concentrations is, of necessity, rather awkward. To overcome this, Sørenson in 1909 proposed a more convenient method of expressing small concentrations of hydrogen or hydroxyl ions. In this method, the $H^+$ or $OH^-$ concentration were written as their _negative_ logarithms. The reason the negative logarithm was chosen by Sørenson is that the most frequently encountered concentrations are lower than unity. For such concentrations, the negative logarithm gives a positive number. However, the student must always be aware of the fact that concentration changes and the corresponding changes in the negative logarithms of these concentrations have the opposite sense. That is, a decrease in the pH corresponds to an increase in $H^+$.

$$- \log [H^+] = pH$$

and

$$- \log [OH^-] = pOH$$

In a number of later instances, this "p-notation" will be used to indicate the negative logarithm of the term that is preceded by p. For example, $pK_a = - \log K_a$

$$pM = - \log [M]$$

The p-notation can be used in writing logarithmic expressions of equilibrium constants. For example, the autoprotolysis constant for water is:

$$[H^+][OH^-] = K_w \rightarrow [H^+][OH^-] = K_w$$

or

$$\log K_w = \log[H^+] + \log[OH^-]$$

Hence,

$$pK_w = pH + pOH \qquad (4\text{-}12)$$

## Expressing Concentrations By Means of Fractions of Dissociation, $\alpha$

A primary strategy in dealing with equilibrium calculations of all kinds is to describe all concentration variables as the product of two factors, i.e., as $\alpha C$. The value of this strategy is that the concentration variable is composed of two separate components, one dependent solely on the analytical concentration, C, not on equilibrium considerations, and the other, $\alpha$, not on analytical concentration (provided the ionic strength and, hence, the activity coefficients remain constant throughout the study) but on equilibrium considerations.

The $\alpha$ values represent the ratios of concentration of individual species, e.g., [HA], to the total concentration of all species of a component, C, i.e., [HA]/C, and are referred to as the dissociation fractions. The sum of the fractions for any system is unity, or $\sum_i^n \alpha_i = 1$. Thus, for a monoprotic acid, $\alpha_o + \alpha_1 = 1$, and for a diprotic acid, $\alpha_o + \alpha_1 + \alpha_2 = 1$.

## Derivation of $\alpha$ Expressions

In a solution of C moles per liter of a weak monoprotic acid, HA, which dissociates in solution according to the equation

$$HA + H_2O \rightleftharpoons H_3O^+ + A^-$$

there are two equations which serve to describe the composition as a function of the acidity and total concentration. These are:

(1) the acid dissociation constant expression,

$$K_a = \frac{[H^+][A^-]}{[HA]} \qquad (4\text{-}13)$$

and (2) the description of the mass balance

$$C = [HA] + [A] = \alpha_o C + \alpha_1 C \qquad (4\text{-}14)$$

where the subscripts $_o$ and $_1$ identify the $\alpha$ in terms of the number of protons lost by the acid. Equations 4-13 and 4-14 may be solved for [HA] and [A] to give

$$[HA] = C\frac{[H^+]}{[H^+] + K_a}; \qquad [A^-] = C\frac{K_a}{[H^+] + K_a} \qquad (4\text{-}15)$$

From Equation 4-15, the $\alpha$ values for a weak monoprotic acid are seen to be

$$\alpha_o = \frac{[H^+]}{[H^+] + K_a}; \qquad \alpha_1 = \frac{K_a}{[H^+] + K_a} \qquad (4\text{-}16)$$

Let us now derive the $\alpha$ values for a diprotic acid, e.g., $H_2S$. Here the dissociation equilibrium expressions are

$$K_1 = \frac{[H^+][HS^-]}{[H_2S]}; \qquad K_2 = \frac{[H^+][S^-]}{[HS^-]} \qquad (4\text{-}17)$$

and the equations based on the $\alpha$ notation, where

$$C = [H_2S] + [HS^-] + [S^-] = \alpha_o C + \alpha_1 C + \alpha_2 C \qquad (4\text{-}18)$$

Hence,

$$\alpha_o = \frac{[H_2S]}{C}, \quad \alpha_1 = \frac{[HS^-]}{C}, \quad \alpha_2 = \frac{[S^-]}{C} \qquad (4\text{-}19)$$

From these, we obtain the following equations

$$K_1 = \frac{[H^+]\alpha_1}{\alpha_o} \quad ; \quad K_2 = \frac{[H^+]\alpha_2}{\alpha_1} \qquad (4\text{-}20)$$

or, canceling out C and rearranging,

$$\alpha_1 = \alpha_o \frac{K_1}{[H^+]} \quad ; \quad \alpha_2 = \alpha_1 \frac{K_2}{[H^+]} = \alpha_o \frac{K_1 K_2}{[H^+]^2} \qquad (4\text{-}21)$$

Recognizing that $\sum_i^n \alpha_i = 1$,

$$1 = \alpha_o + \alpha_o \frac{K_1}{[H^+]} + \alpha_o \frac{K_1 K_2}{[H^+]^2} \qquad (4\text{-}22)$$

which can be rearranged to

$$\alpha_o = \frac{[H^+]^2}{[H^+]^2 + K_1[H^+] + K_1 K_2} \qquad (4\text{-}23)$$

The values of $\alpha_1$ and $\alpha_2$ can now be obtained using Equation 4-24

$$\alpha_1 = \frac{K_1[H^+]}{[H^+]^2 + K_1[H^+] + K_1 K_2} \qquad (4\text{-}24)$$

and

$$\alpha_2 = \frac{K_1 K_2}{[H^+]^2 + K_1[H^+] + K_1 K_2} \qquad (4\text{-}25)$$

It is essential in learning this material to keep it from being just a set of algebraic expressions. Remember that each term in an $\alpha$ expression can be shown to be proportional to a concentration variable. Notice, for example, that in Equation 4-16, both fractions have the same two-term denominator because they represent the concentrations of the two possible species, HA and A. Note also that the concentration of HA is proportional to $[H^+]$, and that of $A^-$ to $K_a$. Similarly, the denominators in Equations 4-22, 4-23, and 4-24 are identical. This forms the basis of a systematic set of rules that can be used to write $\alpha$ values without having to derive them for each new problem that will also reinforce the connections between the chemical and mathematical aspects of equilibrium calculations.

Rules for writing $\alpha$ values involve three simple steps:

(a) the denominator for each $\alpha$ in a component system that we will designate as $H_n B$, is identical; write it first.

(b) the denominator is a decreasing power series in $[H^+]$ (which we will write from now on as simply H), starting with $H^n$, where n is the total number of protons that can dissociate and stopping at $H^0$, or 1. In each successive term, an additional stepwise dissociation constant becomes a factor. There will be a total of $(n+1)$ terms, in which the last is $K_1 K_2 ... K_n$.

(c) Since each term in the denominator is proportional to the concentration of a particular species, the first term, $H^n$, forms the numerator of $\alpha_0$, $H^n$ - 1 $K_1$ for $\alpha_1$, etc. so that, for example, $\alpha_n$ has $K_1 K_2 ... K_n$ as its numerator.

As an example, let us find the expression for the concentration of $PO_4^{3-}$ in a C moles per liter solution. Since this species has lost three protons, the $\alpha$ will be designated $\alpha_3$, and $[PO_4^{3-}] = \alpha_3$ C. The denominator of $\alpha$ starts with $H^3$ since phosphoric acid is triprotic; the term proportional to $[PO_4^{3-}]$ is $K_1 K_2 K_3$ and, therefore, is the numerator.

$$[PO_4^{3-}] = \alpha_3 C = \frac{K_1 K_2 K_3}{H^3 + K_1 H^2 + K_1 K_2 H + K_1 K_2 K_3} \cdot C \qquad (4\text{-}26)$$

If it were $[HPO_4^{2-}]$ being evaluated, then for $\alpha_2 C$, $K_1 K_2 HC$ would be the numerator.

## Proton Balance Equations

The method for carrying out any pH calculation centers around the proton balance equation (PBE). This equation permits you to emphasize the chemical aspects of the problem before the introduction of complex mathematical terms. It, together with all of the equilibrium and balance ($\sum_i \alpha_i = 1$) equations, satisfies the general requirement that, in order to solve a problem with n unknowns, n independent equations are needed.

The PBE matches the concentrations of species which have released protons with those which have consumed protons.

i.e., # protons consumed = # protons released.

Let us first consider the PBE for $H_2O$, not only because it is the simplest, but also since it will be involved in all other PBEs as well.

$$[H^+] = [OH^-]$$

The reason the hydroxide ion concentration measures the concentration of protons released is that when water acts as an acid, i.e., a proton releasing species, it produces a hydroxide ion for every proton released. When water acts as a base, it forms one $H_3O^+$ for every proton consumed. The hydronium ion concentration abbreviated as $[H^+]$ is a measure of the proton consumption of water.

To obtain the PBE for any aqueous solution, the PBE for water is used as a starting point since water is always present. In a solution of a monoprotic acid, HX (strong or weak), the dissociation of HX releases one proton which is measured by the $X^-$ formed in the same process. Thus, the PBE for HX is:

$$[H^+] = [OH^-] + [X^-]$$

In a solution of a strong base such as KOH, the PBE is:

$$[H^+] + [K^+] = [OH^-]$$

Using the neutral KOH as our starting point, its dissociation releases an amount of $K^+$ equivalent to the $OH^-$. The $[K^+]$ is a measure of the $[OH^-]$ released and is therefore equivalent to the protons consumed in this dissociation. Of course, there are many bases which do not directly release hydroxide ions (Brønsted bases). In the case of $NH_3$, for example, protons are consumed to form an equal number of $NH_4^+$ ions. Hence, the PBE for aqueous $NH_3$ is:

$$[H^+] + [NH_4^+] = [OH^-]$$

Let us next consider the case of strong electrolytes MX (where $M^+$ is a cation other than $H^+$ and $X^-$ is an anion other than $OH^-$, since these cases have already been considered above). Considering the neutral solution of MX as a starting point, the dissociation into $M^+$ and $X^-$ neither consumes nor releases protons. One or both of these ions, however, may subsequently consume or release protons. Thus, a solution of a salt such as NaCl will have the same proton balance equation as that of water. In $NH_4Cl$, however, the $NH_4^+$ ion is an acid whose proton release is measured by $[NH_3]$, but the $Cl^-$ does not act as a base in water. Therefore, the proton balance equation for $NH_4Cl$ is:

$$[H^+] = [NH_3] + [OH^-]$$

Similarly, for NaOAc the PBE is:

$$[H^+] + [HOAc] = [OH^-]$$

A case of a strong electrolyte in which both ions are involved in proton balance is that of $NH_4OAc$. Here the PBE is:

$$[H^+] + [HOAc] = [NH_3] + [OH^-].$$

This principle applies equally well to polyprotic acids and bases. In these cases, the concentration terms in the PBE are multiplied by the number of protons consumed or released in the formation of the species in question from the starting material.

The PBE for $H_2S$ is:

$$[H^+] = [OH^-] + [HS^-] + 2[S^{2-}]$$

The PBE for NaHS is:

$$[H^+] + [H_2S] = [OH^-] + [S^{2-}]$$

The PBE for $Na_2S$ is:

$$[H^+] + [HS^-] + 2[H_2S] = [OH^-]$$

## Proton Balance Equations for Mixtures

In the preceding section, PBEs were developed for aqueous solutions having a single analyte.  In reality, since we always included the contribution from water, $[H^+] = [OH^-]$, each PBE was that of a mixture. Continuing this approach, when two or more analytes are involved, simply add the contributions from each to obtain the final equation.  Thus, for a solution that is 0.1 M HOAc and 0.01 M HCOOH(Formic acid), the PBE is:

$$[H^+] = [OH^-] + [OAc^-] + [HCOO^-]$$

Generalizing, the PBE for a mixture of several weak monoprotic acids is:

$$[H^+] = [OH^-] + [B_1^-] + [B_2^-] + \cdots$$

When the mixture includes an acid-base/conjugate pair such as $C_A$ M HOBz (Benzoic acid) and $C_B$ M NaOBz, rather than simply the general expression of [HOBz] or [OBz^-], substitute the equivalent $\alpha C$ product to avoid confusion.  Thus, the PBE is

$$\alpha_0 C_B + [H^+] = [OH^-] + \alpha_1 C_A$$

## Representation of Concentrations of Species by EquiligrapHs

As has been shown, it is relatively easy to write algebraic expressions to represent concentrations of all species of interest ($\alpha C$), even when these expressions seem fairly complex.  In order to visualize how these quantities vary with pH, it is useful to employ graphs, which may be called EquiligrapHs, having Log concentration along the vertical (Y) axis vs. pH along the horizontal (X) axis.

Expressing concentrations in logarithmic terms is important from several standpoints:  (1) since pH (defined as-log [H]) is itself a logarithmic function, it is appropriate to express the concentrations of other species in the same manner and (2) these logarithmic plots are simple, consisting of straight-line segments, connected smoothly in the regions where pH values are close($\sim \pm 1$) to pK values.  This simplicity results from the large differences in successive pK values which results in the predominance of a single species when the pH is 1.0 or more units away from a pK value.  The slopes ($\delta\log \alpha/\delta pH$) of the line segments have integral values, 0, 1, 2, etc., depending on the number of protons gained or lost in the transformation of the species in question to the predominant species at the pH in question.

Thus, the $\log \alpha_3$ (for $PO_4^{3-}$) rises with increasing pH with a slope of zero in the pH range where $PO_4^{3-}$ predominates, of unity where $HPO_4^{2-}$ is the major component and of three where $H_3PO_4$ predominates.  The concentration of any component can be readily obtained from these plots by the simple addition of two terms.  For example, since $[CO_3^{2-}] = \alpha_2 C$, $\log [CO_3^{2-}] = \log \alpha_2 + \log C$.

Please note the great ease with which such log C vs. pH diagrams can be drawn.  Figure 4-1, representing a 0.02 M solution of $H_2CO_3$, was drawn according to the following generally applicable instructions.

(1)     Draw and label a 14 x 14 square on coordinate paper and the diagonals representing $[H^+]$ and $[OH^-]$.  The diagram, so far, is common to all systems.

(2)     Draw a horizontal line at -log C = -log 0.02 = 1.70.  This is called the **system line**.

(3)     At pH values corresponding to $pK_1$ and $pK_2$ for $H_2CO_3$, **corrected for activity effects at the ionic strength of the solution (see Chapter 3)** (6.37 and 10.32, respectively), mark points on the system line.

(4)     At these points, draw line segments of slopes plus and minus unity (angles of $\pm 45°$) which extend down until they either reach the boundaries of the square or the pH corresponding to the next pK value.  In the latter case, the line segment is extended further, but with twice the slope.  (In the case of a triprotic acid, if the line segment later reaches a pH value corresponding to yet another pK value, the slope now changes to the next larger integer.

Both the size and the sign of the slope will reflect the number of protons gained (slope is +) to transform the species represented by the line you are drawing to the **predominant** species.  Thus, the $[CO_3^{2-}]$ line has a slope of +2 where $H_2CO_3$ predominates (roughly pH 0 to 5), of +1 from pH 7 to 10 where $HCO_3^-$ predominates, and zero at pH values above ~11, the region where $CO_3^{2-}$ is the major species.

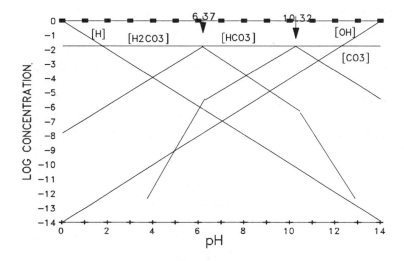

**Figure 4.1 Log C vs. pH Diagram: 0.02 M Carbonate**

From such charts, it is possible (1) to show how the concentrations of various species in acids and bases change with pH and to indicate which species are important and which have negligible concentration at a given pH value, and, as will be seen below,  (2) to calculate the pH values of such solutions quickly and accurately with the help of the PBE, described below.

It must not be thought that plotting the $\alpha$ values rather than log $\alpha$ values is useless.  Even though it is very difficult to draw this kind of graph properly freehand, it can be done using the spreadsheet, just as easily as the other (see Problem 4-3).

## Graphical-Algebraic Solution
## of the Proton Balance Equation

The method for carrying out any pH calculation centers around the PBE. This equation permits you to emphasize the chemical aspects of the problem before the introduction of complex mathematical terms. Once the PBE is written, the graph describing all the components of the system is drawn. For a particular problem, attention will be focussed on the intersections of those species on the left-hand side of the PBE with those species on the right-hand side. The intersection of the two species lines at the highest value of the concentration (the intersection closest to the top of the diagram) is called the **principal intersection**, and gives the approximate pH of the solution.

The approximate graphical solution is used to simplify the PBE. All terms that are seen to be less than 5% of the main components in the PBE are discarded. The choice of 5% as a cutoff criterion is based on the usual limit of $\pm 0.02$ units accuracy in pH measurements. Since pH is logarithmic, the corresponding numerical difference is $10^{0.02}$, which is 1.05, or 5%. The difference in logarithms corresponding to 5% (or 1 part in 20) is -1.30. Therefore, a vertical distance of 1.3 log units, or more, **measured at the pH selected by the principal intersection** on the graph, serves to identify terms that may be discarded (of course, the criterion of $\pm 0.02$ may be altered to suit the particular experimental situation; for $\pm 0.01$, show that a length of -1.7 applies and for $\pm 0.05$, a length of -1.0).

For the remaining terms, the appropriate $\alpha C$ expression is substituted for every species except [OH⁻], which becomes $K_w/[H^+]$. Now the equation is further simplified with the help of the graph by discarding terms in the denominators of the various $\alpha$ expressions, corresponding to species of negligible concentration ($<5\%$). The resulting equation now leads to an appropriate algebraic expression for the pH of the solution that is reliable to $\pm 0.02$. In most cases, including rather difficult problems, the final algebraic expression is a simple equation. This method has another advantage in dealing with those unusual problems in which the final algebraic expressions are not so simple. Not only are derivations of such expressions straightforward, but the components responsible for the complexity can be clearly identified. The following examples illustrate the method.

Returning again to the EquiligrapH of 0.02 M $H_2CO_3$, we will learn that in addition to a value of the concentration of each species as a function of the pH, the graph may be used  to calculate the pH values of this and other solutions by means of the PBE.

The PBE for 0.02 M $H_2CO_3$ is:

$$[H^+] = [OH^-] + [HCO_3^-] + 2[CO_3^{2-}]$$

For this problem, there are three intersections to examine; that of $[H^+]$ with (a) $[OH^-]$, (b) $[HCO_3^-]$, and (c) $[CO_3^{2-}]$.  From the diagram it can be seen that (b) takes place at the highest concentration and is therefore the Principal Intersection (P.I.), which gives the approximate pH.   Drawing a vertical line at this pH, it is seen that this crosses those of $[OH^-]$ and $[CO_3^{2-}]$ far below  the P.I., at more than 1.3 log units.  This allows us to simplify the PBE to:

$$[H^+] = [HCO_3^-]$$

Now we can use $\alpha C$ to obtain the expression which has only one variable, namely $[H^+]$.

$$H = \frac{0.02\ HK_1}{H^2 + K_1\ H + K_1\ K_2} \tag{4-27}$$

Consulting the EquiligrapH once again, it can be seen that at the P.I., $[H_2CO_3]$ is at least 1.3 log units above the other two carbonate species concentrations.  Hence, $H^2 > HK_1 > K_1K_2$ and

$$H = 0.02K_1/H \quad \text{or} \quad H^2 = 0.02K_1$$

Converting this to its logarithmic form, we obtain

$$pH = 1/2(pK_1 - \log 0.02)$$

from which we may deduce that the general equation giving **the pH of a weak diprotic acid is $1/2(pK - \log C_A)$**.  It will be instructive for you to compare this with one derived for the pH of a weak, monoprotic acid. Why are they identical?

As long as we went to the trouble of drawing the EquiligrapH for 0.02 M $H_2CO_3$, let us use it for two other problems, namely the pH values in 0.02 M solutions of $NaHCO_3$ and $Na_2CO_3$.

NaHCO$_3$ PBE: $[H_2CO_3] + [H^+] = [OH^-] + [CO_3^{2-}]$

P.I.:

$$[H_2CO_3] = [CO_3^{2-}]$$

Using $\alpha C$:

$$CH^2 = CK_1K_2$$

or

$$pH = 1/2(pK_1 + pK_2)$$

Na$_2$CO$_3$ PBE: $2[H_2CO_3] + [HCO_3^-] + [H^+] = [OH^-]$

P.I.:

$$[HCO_3^-] = [OH^-]$$

Using $\alpha C$:

$$CK_1H/(H^2 + K_1H + K_1K_2) = K_{w/H}$$

or

$$pH = 1/2(pK_w + pK_2 + \log C)$$

It is useful to notice that these last two pH equations represent the simplest appropriate ones for an amphiprotic salt and a weak base, respectively.

## Acid-Base Equilibrium Calculations with the Spreadsheet

Now that we have described how to draw A/B diagrams freehand, let us see how use of a spreadsheet program can simplify the task. We will start with a generally useful spreadsheet we will call **A/B START**. A typical acid-base situation can serve as illustration. At the computer, call up your spreadsheet program. When the spreadsheet layout appears on the screen, let us first suitably label all the columns with the quantities needed in the problem, using the first row, starting with the critical variable, pH in cell **A1**. Now, in the adjacent columns, enter labels for these related and pertinent concentration variables: in **B1** [H], in **C1** [OH], and in **D1** log[H], in **E1** log[OH].

Next, enter (starting in row 2) in the column labeled pH, a range of values of the critical variable, say 0 to 14, at suitable intervals, e.g., increments of 1. This is accomplished by calling for the menu (keyboard symbol: /) and then using the BLOCK FILL function by appropriate moves of the cursor or using the keyboard symbols: /EF{/BF;/DF} for range A2..A16. (Note that if you wanted more closely spaced intervals, e.g., 0.1, you would simply call /EF for the block A2..A141, with the same ease). Move now, still in row 2, through the remaining columns, entering the formula (starting with: +) that relates the particular variable to the critical variable and subsequently to whatever variable or combination of variables that have already been defined by entries in earlier columns. The formula appears at the lower left corner of the screen, but the numerical value appears in the cell. For this problem, in B2, enter the formula for [H] from pH, that is, $+10^\wedge$-A2, in C2, for [OH], enter $+10^\wedge$-14/B2. (The symbol: $^\wedge$ signifies that the next number or cell descriptor is an exponent.)

To prepare for the Log Concentration vs. pH graph, or EquiligrapH, in columns D and E, we will enter the log [H] and log [OH] values, i.e., in D2, enter @log(B2), and in E2, @log(C2). (The ampersand, @, introduces specially designed functions). It will be instructive to pause to develop this part of the problem further. Use the BLOCK COPY function to enter the values for [H], [OH], log[H], and log[OH] (from the keyboard: /EC{/BC; /C}) and sources B2..E2, destined for B3..E16. All of these values are calculated and appear instantly on the screen. To view your work, let us construct a graph, /G{/G; /G} of the XY type (selected from menu by calling : /GGX from the keyboard). As X values, enter A2..A16. In the series of Y values, enter 1st Series: D2..D16; 2nd Series: E2..E16., then View (Figure 4-3). This is the framework of all subsequent EquiligrapH diagrams for A/B problems. (This basic layout might be worth saving, /FS{/FS; /FS}, giving it the name we choose, e.g., A/BSTART, and then retrieving it as A/BSTART.WQ1{.WKQ; .WK1} whenever needed as a jumping off place for most other A/B problems. We simply enter the additional appropriate columns of relevant material, and store the resulting specific system under another, appropriate name. Let us do that now for a monoprotic system, HB.

The next step in our exercise is to develop the expressions for [HB] and [B], the species concentrations for the monoprotic acid, HB, and its conjugate base, B. These, like all other concentrations of species of interest, are written as the product of two factors, $\alpha C$, as described above.

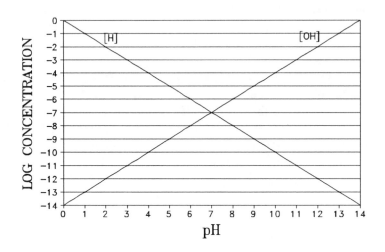

**Figure 4.2 Basic EquiligrapH**

## Example 4.1

Let us use HOAc as our example. Write labels in F1..I1 for [HOAc], [OAc⁻], and their logarithms. In cell F2, enter the formula for [HOAc] that would apply to a 0.1 M acetic acid (pK = 4.74): +0.1*B2/(B2 + 10^-4.74); in G2: +0.1-F2 since the sum of [HB] and [B] must be C, or in this case, 0.1. As before, the log values come next: in H2: @log(F2); in I2: @log(G2). Complete the spreadsheet by using /EC{/BC; /C} with source F2..I2, destination F3..I16.

Return to the graph by /GS{/GS; /G} so that you may designate the 3rd series {C series in 123} by H2..H16 and the 4th {D in 123} by I2..I16. View the graph which is a display of the Equiligraph of a 0.1 M acetic acid solution (Figure 4-3). Here the curves are drawn for the logarithms of [H⁺], [OH⁻], [HOAc], and [OAc⁻]. The additional Pointer curves will be described below. (NOTE: All the graphs you construct from one spreadsheet can be stored, by using the command /GNC{/GNS; /GNC}. Each of the graphs can be displayed by choosing the unique name you give it. This will permit you to fully exploit the data in one spreadsheet. Remember to store the changes before leaving the file: /FAR{/FSR; /FS}.

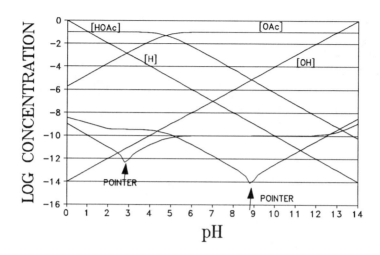

**Figure 4.3  0.1 M Acetic Acid or Sodium Acetate System**

## Example 4.2

Construct an equiligrapH for the 0.02 M carbonate system, i.e., one that applies to 0.02 M solutions of either $H_2CO_3$, $NaHCO_3$, or $Na_2CO_3$. Find the pH of each of these solutions with the help of the pointer function. Neglecting the effect of ionic strength on the $pK_a$ and $pK_{a2}$ values, the same diagram will apply to all three. Of course, the PBEs will differ as will, therefore, the pointer functions needed to obtain the pH values of the three solutions.

First, retrieve /FR{/FR; /FR} the **A/BSTART** file, and starting in **F1**, complete labeling columns as $[H_2CO_3]$, $[HCO_3^-]$, $[CO_3^{2-}]$, their logs, and finally the three pointer functions for the three solutions.

In cell **F2**, use the formula for $[H_2CO_3]$, i.e., 0.02*B2^2/(B2^2 + 10^-6.13*B2 + 10^-(6.13 + 10.33)); in **L2**, the pointer for $H_2CO_3$: +C2+G2+2*H2-B2.

When all of the cells **F2..N2** are prepared, the copy command /EC{/BC; /C} from **F3..N16** will complete the spreadsheet. Construct the graph using pH for the X series and as the six Y series log[$H_2CO_3$], log[$HCO_3^-$], log[$CO_3^{2-}$], and the three pointers. (If QPRO is used, the annotate function will permit you to conveniently draw the two lines corresponding to log[H] and log[OH].) Figure 4.4 represents the graph just described.

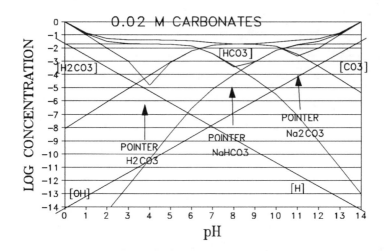

**Figure 4.4 0.02 M Carbonate System**

This procedure can be utilized to develop EquiligrapHs for all kinds of acid-base systems. In fact, by utilizing absolute addresses, or Block Names for adjustable parameters of concentration and the pK values, it is possible to solve a large number of problems with a single spreadsheet. Remember to transfer the columns of values of those problems you wish to preserve before changing to the next set of adjustable parameters. This is accomplished by the command /EV{/BAV; /RV}.

Even in those cases where the solution does not reduce to the pH value of the principal intersection, making this value increasingly uncertain, one can see from the EquiligrapH and the PBE both the source and extent of the error. It must be remembered, however, that the PBE itself can guide us to the rigorously correct solutions for such problems, regardless of their complexity. This can be done with the spreadsheet, and a novel way to use the PBE called the POINTER function.

### The Pointer Function

As we saw in Chapter 1, even very complex algebraic expressions can be solved by a simple graphical method involving the POINTER function.

Application of this idea to acid-base as well as many other analytical calculations is far from complicated. This is especially true since for pH, pM, etc., no matter how complex the expression may look, we know that any single aqueous solution can have only one pH value. Hence, the expression has a **single**, unique solution. Further, when the problem involves pH calculations, the range of values to be examined is obviously between 0 and 14 in almost every case.

For purposes of illustration, let us use a 0.10 M $(NH_4)_2HPO_4$, for which the PBE is:

$$2[H_3PO_4] + [H_2PO_4^-] + [H^+] = [OH^-] + [NH_3] + [PO_4^{3-}]$$

This can be rearranged to give the POINTER function, P:

$$P = 2[H_3PO_4] + [H_2PO_4^-] + [H^+] - [OH^-] - [PO_4^{3-}] - [NH_3]$$

The value of P will vary with pH but will be uniquely ZERO at the pH of the solution. While this could be represented by the intersection of the P function with pH axis, it can be conveniently transformed to the log of

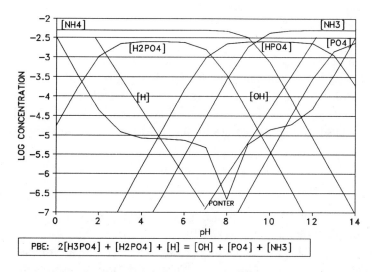

Figure 4.5 Finding the pH in 0.0025 M $(NH_4)_2HPO_4$

its absolute (one cannot obtain the log of a negative number) value, or @LOG(@ABSP). This is seen in Figure 4.4 by the minimum in the pointer function which points to pH = 8.0, the answer. Pointer functions obtained in the same manner have been included in Figures 4.2 and 4.3.

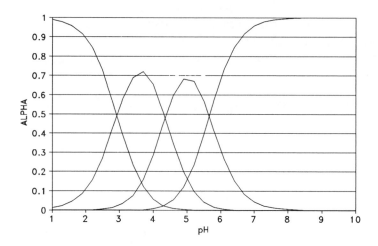

**Figure 4.6 Distribution of Citric Acid Species as a Function of pH**

In these or any similar problems, remember that significant improvement in definition of the solution may be obtained by adding more pH values near that which is the answer (say, within ±1), but in much smaller increments (say, 0.1 or 0.05); follow this by /EC{/BC; /C} and then sorting the entire set of pH values by the database command /DSB{/DSB; /DSR}.

## PROBLEMS

4-1     The following equations are the simplest and most frequently used for pH calculations. Write PBE's for these cases and verify the validity of the equations using the graphical method for simplifying the problems. What are the simplifying assumptions made and what are the conditions under which they do not apply?

(a) strong acid: $pH = -Log\ C_A$
(b) strong base: $pH = pK_w + Log\ C_B$
(c) weak acid: $pH = 1/2(pK_a - Log\ C_A)$
(d) weak base: $pH = 1/2(pK_w - pK_a + Log\ C_B)$
(e) amphiprotic salt: $pH = 1/2(pK_{a1} + pK_{a2})$
(f) Buffer equation: $pH = pK_a + Log\ (C_B/C_A)$
(g) Mixture of weak acids: $pH = \sum_i (C_{ai} K_{ai})$

**4-2**    Develop the spreadsheets and corresponding Equiligraphs for examples involving equations (a) to (f) above, using the Pointer function to find the pH for each case. Initially, use pH increments of 1. When you have located the pH $\pm$ 0.3, return to the spreadsheet, and add an additional set of pH values in increments of 0.01 centered on the preliminary pH value. Use Block Copy and Sort to put the values in order on the spreadsheet. To obtain the value of the pH, change Scale from Automatic to Manual with a High of ($pH_{prelim}$ + 0.15) and a low of ($pH_{prelim}$ -0.15) and an increment (on GraphMenu) of 0.02.

**4-3**    Using the pK values listed in Appendix C for various ligands which are also Brønsted bases, develop spreadsheets and EquiligrapHs for a number of common ligands. Preserve these for use in conjunction with questions concerning the role of pH in metal complex formation that will be introduced in Chapter 5.

**4-4**    Develop the spreadsheet and EquiligrapHs for an "acid rain" sample that contains $3 \times 10^{-5}$ M $H_2CO_3$, $1 \times 10^{-4}$ M $H_2SO_4$, and $2.5 \times 10^{-5}$ M $HNO_3$. What is the pH of "acid rain"?

**4-5**    Develop the spreadsheet and EquiligrapHs for an "industrial waste effluent" that has $5 \times 10^{-3}$ M phenol, $5 \times 10^{-5}$ M HCN, and $1.2 \times 10^{-5}$ M $H_2S$. What is the pH of this effluent?

**4-6**    Using citric acid ($pK_{a1}$ = 2.931, $pK_{a2}$=4.361, $pK_{a3}$=5.667) as an example, prepare a spreadsheet enabling you to prepare two graphs, one in which $\alpha$ values are plotted against pH, as in Figure 4.5, and the other in which the Log $\alpha$ values vs. pH appear. Which is more useful for pH calculations, and why? Would your answer be the same with other uses?

**4-7**    What is the value of K to be used in the following examples? (Use
         Davies' Equation 3-10)

        (a)  0.01 M HOAc in 0.15 M NaCl.  p*K = 4.756
        (b)  0.01 M HOAc (use successive approximation)
        (c)  0.10 M $NH_4Cl$.  p*K = 9.244
        (d)  0.005 M $H_3PO_4$ in 0.10 M KCl.  p*$K_1$ = 2.148,
            p*K2 = 7.199, p*$K_3$ = 12.35

# Chapter 5

# Metal Complex Equilibria

---

The formation of metal coordination complexes occurs when a metallic cation, which is an electron-pair deficient polybasic Lewis acid, reacts with several Lewis basic species, or **ligands**, whose number is related to the coordination number of the metal ion. The **coordination number** of metal ion refers to the number of bonding atoms in the ligands that can arrange themselves around the central metal ion, and depends largely on the size of the metal ion as well as the size of the coordinating groups. The ligands are usually either neutral, as in the case of $NH_3$ or $H_2O$, or negatively charged, as with $CN^-$, $OH^-$, or halide ions. Ligands are classified by the number of atoms capable of bonding to atoms; those with one are called **monodentate** or **simple** ligands; those with two or more, **polydentate** or **chelating** ligands.

The strongest Brønsted acid in aqueous solution is the hydrated proton or hydronium ion $H_3O^+$ because the proton is not free, but hydrated in aqueous solutions. Analogously, the hydrated form of a metal ion will be the strongest Lewis acid of any of the complexes of this metal ion in aqueous media. Therefore, the acid-base chemistry of metal complexes is complicated by the existence of a large number of reference acids, consisting of each of the hydrated metal ions, which is in contrast to the simplicity achieved in the Brønsted A/B system, where there is but one reference acid, the hydronium ion. We will follow the custom of usually referring to the hydrated metal ion by the symbol, $M^{n+}$, analogous to $H^+$, rather than include $H_2O$ in order to simplify equation writing.

In simple coordination complexes, metal ions combine with monofunctional (monodentate) ligands in a number equal to their coordination number, e.g., $Ni(NH_3)_6^{2+}$, $Fe(CN)_6^{4-}$, $FeCl_4^-$, and $Co(H_2O)Cl_5^{2-}$. In chelate complexes, metal ions interact with polyfunctional ligands, each capable of occupying more than one position in the coordination sphere of the metal. The charge of a metal complex, being the algebraic sum of the electronic charges of the metal ions and the ligands, may be positive, e.g., $Zn(H_2O)_6^{2+}$, neutral, e.g., $GeCl_4$, or negative, e.g., $CoCl_4^{2-}$. Chelates may also be cationic ($Cuphen_2^+$) neutral ($Fe(Ox)_3$), or anionic ($CaEDTA^{2-}$).

Metal complexes are very useful in analytical determinations and separations. Both positively and negatively charged complexes including chelates are useful as **masking agents** and, particularly in the case of chelates, as **titrants**. Masking refers to the reduction of the concentration of a hydrated metal ion to a point at which the metal ion does not significantly participate in a reaction of interest, by virtue of formation of a sufficiently stable water-soluble complex.      For example, the addition of $NH_3$ to an $Ag^+$ solution will prevent the precipitation of AgCl that otherwise would occur when $Cl^-$ is added. If $I^-$ were added, however, the presence of $NH_3$ would not prevent the precipitation of AgI. From this we can say $NH_3$ effectively masks $Ag^+$ in the presence of $Cl^-$ but not in the presence of $I^-$.   Cyanide ion forms a sufficiently stable, charged Ag complex, $Ag(CN)_2^-$, to effectively mask $Ag^+$ in the presence of I.   $Zn^{2+}$ can be masked by EDTA so that no ZnS can be precipitated. Chelating titrants for complexometric methods require that very stable 1:1 charged complexes are formed.

Neutral chelates are characterized by a low solubility in water, a significant solubility in organic solvents, and, in some cases, volatility are widely used in metal separation processes, such as precipitation, extraction, or gas chromatography.

Analytical applications of all types of metal complexes will be discussed later in this book.

In general, the overall formation constants of chelates are higher than those of similar simple coordination complexes. Even if the constants were the same, however, there is a tremendous difference in the extent of complexation that can be best explained by an example. Divalent copper ion, having a coordination number 4, can react with four monodentate ligands, L, or one quadridentate chelating agent, $L'$. Let us assume that the overall formation constant in each case is $10^{12}$ and that the equilibrium concentration of ligand in each case is $10^{-3}$ M. Then,

$$\frac{[ML_4]}{[M][L]^4} - 10^{12} \text{ and } \frac{[ML']}{[M][L']} - 10^{12}$$

From these equations, it may be seen that the concentration ratio of $[CuL_4]:[Cu]$ is unity in the case of the monodentate ligand, but that $[CuL']:[Cu]$ is $10^9$, demonstrating the greater power of the chelate to reduce the "free", i.e., hydrated, $Cu^{2+}$ concentration even when the equilibrium formation constant is the same as for the simple ligand. The formation constants of similar simple and chelate complexes are dramatically different; chelates are inherently more stable. The bond energies of, for example, Cu-N bonds are not much different when the N is part of the $NH_3$ molecule or the $H_2NCH_2CH_2NH_2$(n) or $H_2NCH_2COOH$ (glycine) molecules. The difference in $\Delta G°$, called the chelate effect, is due to the greater increase in a fundamental thermodynamic quantity called the entropy.

The treatment of metal-complex equilibrium calculations follows the same pattern developed above (Chapter 4) for proton transfer equilibria. Two relatively minor, but noteworthy, differences are (1) in Brønsted acid-base reactions and metal-complex reactions, their corresponding equilibrium constants are conventionally written as dissociations. With metal complexes, however, reactions and equilibrium constants are written in the reverse manner as formations. The difference in nomenclature or symbols used in describing metal-complex systems must not be allowed to obscure the fundamental similarity of the two A/B systems. (2) Successive formation constants of metal complexes are in general much closer than successive Brønsted acid dissociation constants. This results in calculations in which simplifying assumptions can not be made readily. For example, although it is possible to calculate the hydrogen ion concentration in a solution of $H_2S$ by ignoring the second dissociation constant, it is not possible to calculate the concentration of free $NH_3$ in a solution containing $Cu(NH_3)_2^{2+}$ without taking the several stepwise formation constants into account.

The Lewis acidity level is denoted by pM, the negative logarithm of the concentration of a metal ion M, just as the pH is used as a measure of the Brønsted acidity level. Further, it is necessary to consider stepwise formation constants in metal complex equilibria just as is done with calculations involving polybasic Brønsted acid dissociations. Just as those calculations are systematized and simplified by the use of $\alpha C$ to describe the concentrations of all species, so too is the definition of a set of $\alpha$ values that represent the fractions of the total concentration present as each metal complex species in these cases.

## GENERAL METAL-COMPLEX EQUILIBRIA

A generalized metal complex formation can be represented as follows (charges have been omitted for convenience):

$$M + L \rightleftharpoons ML$$

The equilibrium expression corresponding to this reaction is

$$K_{f_1} = \frac{[ML]}{[M][L]} = \frac{\alpha_{ML} \cdot C_M}{\alpha_M \cdot C_M [L]} = \frac{\alpha_{ML}}{\alpha_M \cdot [L]} \tag{5-1a}$$

which yields a relation between $\alpha$ values and $K$ that is independent of $C$ (just as was seen in Chapter 4).

If more than one ligand is bound to the metal ion, the stepwise formations and their corresponding equilibrium constants can be represented similarly:

$$ML + L \rightleftharpoons ML_2; \quad K_{f_2} = \frac{[ML_2]}{[ML][L]} = \frac{\alpha_{ML_2}}{\alpha_{ML} [L]}$$

$$ML_2 + L \rightleftharpoons ML_3; \quad K_{f_3} = \frac{[ML_3]}{[ML_2][L]} = \frac{\alpha_{ML_3}}{\alpha_{ML_2} [L]} \tag{5-1b}$$

$$ML_3 + L \rightleftharpoons ML_4; \quad K_{f_4} = \frac{[ML_4]}{[ML_3][L]} = \frac{\alpha_{ML_4}}{\alpha_{ML_3} [L]}$$

By combining the equations for the individual reaction steps, an overall metal-complex formation reaction and the overall formation constant expression may be written

$$M + nL \rightleftharpoons ML_n; \quad \beta_n = \frac{[ML_n]}{[M][L]^n} \qquad (5\text{-}2)$$

where

$$\beta_n = K_{f1}K_{f2}K_{f3}\ldots.K_{fn}$$

The symbol $\beta_n$ is referred to as the overall formation constant and is also used for all products of stepwise formation constants. When the overall formation equation is used, we must remember that intermediate complexes, although not mentioned, do play a part. If we did, this would be as incorrect as the assumption that in a solution of $H_2S$ there are 2 moles of $H^+$ for every mole of $S^{2-}$, that is, without taking $HS^-$ into account.

Metal-complex equilibria calculations are accomplished in the same manner as those for polybasic acid equilibria. For this purpose, we have defined a set of fractions $\alpha_m$ to $\alpha_{MLn}$ to represent the ratios of the concentrations of the metal containing species to the analytical concentration of the metal, $C_M$. Thus, incorporating the equilibrium constant expression in Equations 5-1 and 5-2 and rearranging, we obtain in (5-3) expressions for the $\alpha$ values that are functions of the equilibrium constants and the free (i.e., not bound to M) ligand concentration.

$$\alpha_M = \frac{1}{1 + \beta_1[L] + \beta_2[L]^2 + \beta_n[L]^n}$$

$$\alpha_{ML} = \frac{\beta_1[L]}{1 + \beta_1[L] + \beta_2[L]^2 + \beta_3[L]^3 + \beta_4[L]^4} \qquad (5\text{-}3)$$

$$and$$

$$\alpha_{ML_2} = \frac{\beta_2[L]^2}{D}; \; \alpha_{ML_3} = \frac{\beta_3[L]^3}{D} \; ; \; \alpha_{ML_4} = \frac{\beta_4[L]^4}{D}$$

where the $\beta$ values represent the products of stepwise formation constants as shown in Equation 5-2 and the symbol D represents the denominator which is the same for all of the $\alpha_{ML_i}$ in the set.

The simple rules for remembering how to write a equations are quite similar to that used earlier for Brønsted A/B $\alpha$'s (Chapter 4). These differ mainly in that the denominators are an **ascending** power series in [L], not descending powers of [H$^+$].

(1)     The denominator for each $\alpha$ in a component system is identical.
(2)     The denominator is an ascending power series in [L], starting with [L]$^0$ and ending with [L]$^n$, where n is the total number of ligands that can bond with the central metal ion. In each successive term, a new cumulative formation constant, $\beta$, will be a factor. There will be a total of (n + 1) terms, of which the last is $\beta_n$[L]$^n$.
(3)     Since each term in the denominator is proportional to the concentration of a particular species, the first term, [L]$^0$ or 1, forms the numerator of $\alpha_M$, the second for $\alpha_{ML}$, etc.

### Cases Where the Equilibrium Concentration of the Ligand is Known

Finding the concentrations of all metal complex species or, to put it another way, describing the composition as a function of ligand concentration, is simple when the equilibrium concentration of the ligand is known. Three variations can be considered.   First, the equilibrium concentration is given.  Second, the total ligand concentration is so much larger than the total metal concentration that the amount of ligand consumed by complex formation can be ignored.  Third, the ligand concentration is not too much larger than the metal ion concentration, requiring a calculation of the equilibrium ligand concentration.

Once the equilibrium ligand concentration is known, then the $\alpha_i$ values and, therefore, the $\alpha_i C_M$ for all the species can be obtained. This represents an exact analogy to the procedure in finding the concentrations of all the species in a Brønsted acid-base system when the pH is known.

### Example 5.1

Let us consider a solution in which $C_{Cu}$ is 0.005 M and with various, but known, equilibrium concentrations of ammonia.  What are the concentrations of each Cu-containing species at any or all of the equilibrium concentrations of NH$_3$? The stepwise formation constants for copper amine complexes are Log $K_{f1}$ = 4.31, Log $K_{f2}$ = 3.67; Log $K_{f3}$ = 3.04 and Log $K_{f4}$ = 2.30.

The manual construction of these diagrams, however, is more difficult than that for most Brønsted A/B systems because the successive formation constants usually have values that are so close together that in a number of regions significant concentrations of several species are involved, necessitating point by point plotting from calculated $\alpha$ values if use of a spreadsheet program were not available. [When a chelate complex is formed, however, the stepwise constants are farther apart so that, particularly with polydentate ligands, the graphical diagrams can be as easily constructed manually as those for Brønsted acid-base systems (see Figures 5.2 and 5.3 below)].

As with the acid base system, when the spreadsheet layout appears on the screen, use the first row to suitably label all the columns with the quantities needed in the problem starting with the critical variable, log [NH$_3$] in column A. Now, in the adjacent columns, enter labels for these related and pertinent concentration variables: [NH$_3$] , [Cu$^{2+}$] , [Cu(NH$_3$)$^{2+}$],.....[Cu(NH$_3$)$_4^{2+}$] as well as the labels for all the Cu-containing species, log [Cu(NH$_3$)$_i^{2+}$].

Next, enter starting in cell **A2** a range of values of log [NH$_3$] from -6 to 0, at suitable intervals, e.g., increments of 0.1. This range is bounded by a value low enough (at least one unit lower than **-log $K_{f1}$**) so that [Cu$^{2+}$] predominates to one so high (at least one unit higher than **-log $K_{f4}$**) that the fully formed complex is the major species. This is accomplished by the spreadsheet command: /EF{/BF; /DF}. Now, for cell **B2**, write the formula $+10^A$ defining [NH$_3$], in **C2** for [Cu$^{2+}$] the formula for $\alpha_{Cu}C_{Cu}0.005/(1+B2*10^4.31+B2^2*10^(4.31+3.67)+...+B2^4*10^(4.31+3.67+3.04+2.30)$. This value in **C2** is used as the basis for **D2**, i.e., $+C2*10^4.31*B2$, because [Cu(NH$_3$)$^{2+}$] = $\alpha_{Cu}C_{Cu}\cdot\beta_1$[NH$_3$].

Remember that all $\alpha_{ML}$ values have the same denominator and, since $\alpha_{Cu}$ is simply 1/denominator, it can be transformed to the other $\alpha$ values by multiplying by the appropriate numerator $\beta_1$[NH$_3$], $\beta$[NH$_3$]$^2$, etc. Once you have written the correct formulas in **D2** to **G2** (for [Cu(NH$_3$)$_4^{2+}$]) in the next five columns (**H2** to **L2**), write formulas for the logarithms of each of the species, e.g., in **H2**, place @LOG(C2), and for forth. Now, copy the values from **B2** to **L2** by the command /EC{/BC; /C} over the entire range you selected in column A.

As before, all of these values are calculated and appear instantly on the screen. To view your work, construct a graph as described before, with the series corresponding to the log concentration of each of the five Cu-containing species, then View (Figure 5.1). (NOTE: All the graphs you construct from one spreadsheet can be stored, by using the command /GNC{/GNS; /GNC}. This will permit you to fully exploit the data in one spreadsheet. Remember to store the changes before leaving the file: **/FAR-**{/FSR; /FS}.

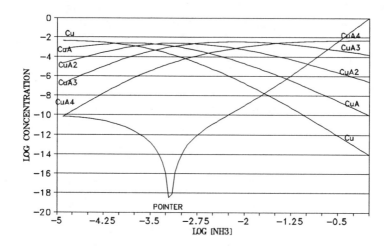

**Figure 5.1. EquiligrapH for 0.005 M Cu²⁺ and NH₃; POINTER for 0.005 M Cu(NH₃)₂²⁺**

Some of the effects of closely spaced constants can be illustrated with the help of Figure 5.1. For example, it will be noticed that the concentrations of the intermediate complexes from $Cu(NH_3)^{2+}$ to $Cu(NH_3)_4^+$ never reach $C_M$, that is, the value of 0.005 M, since there are always significant fractions of the total copper present as other species. It can be shown that unless two successive constants are at least 3.2 log units apart, the intermediate species will not reach a maximum concentration of 0.95 × C.

In analogy with the Brønsted acid-base EquiligrapHs, the intersections of the log C curves for the different species correspond to values of Log $[NH_3]$ that are equal to the log $K_{fi}$ values. Because successive formation constants are so close, however, this does not mean that only two species have significant concentrations at such intersections.

## Example 5.2

Let us consider a similar problem by asking about the concentrations of all Cu-containing species in a solution of 0.005 M $Cu^{2+}$ containing various concentrations of uncomplexed ethylenediamine(en). There are two stepwise formation constants, log $K_{f1}$ = 10.72 and log $K_{f2}$ = 9.31. Constr-uct the spreadsheet as before, but using the range of log [en] from -13 to -7, in steps of 0.1 (Figure 5.2).

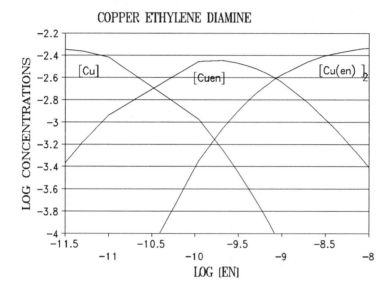

**Figure 5.2　EquiligrapH for 0.005 M $Cu^{2+}$ with Ethylenediamine(en)**

It is instructive to compare the copper-ammonia and the copper-ethylenediamine systems. The greater stability of chelates over simple coordination compounds is dramatically seen in the much lower en concentration, $10^{-11}$ M, as compared to $10^{-5}$ M for $NH_3$, at which the free copper ion concentration begins to fall. It will also be observed that the greater separation of the stepwise constants than is observed with simple coordination complexes results in a simplified diagram that clearly resembles that for a dibasic acid, such as $H_2CO_3$. The improvement in the stability of copper complexes achieved in going from ammonia, a monodentate ligand, to ethylenediamine, a bidentate ligand, is further increased with the quadridentate ligand "trien", trimethylenetetramine, ($NH_2CH_2CH_2NHCH_2CH2NHCH_2CH_2NH_2$). Log $\beta_1$ = 20.4.

**Example 5.3**

Describe the composition of a 0.005 M $Cu^{2+}$ solution containing various trien concentrations. Construct the spreadsheet for this system as for the others, except that the log [trien] range from -21 to -15 is used (Figure 5.3). In this system the concentration of free copper ion begins to drop at a free ligand concentration below $10^{-20}$ M. Since only a 1:1 complex is formed in this case, the diagram is quite simple, as is the analogous HOAc EquiligrapH. Metal-EDTA complexes have diagrams of similar shape.

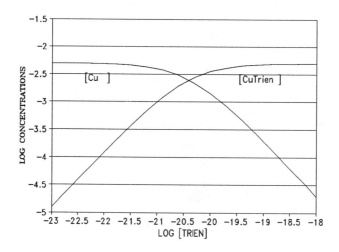

**Figure 5.3  Distribution of Species in Cu - Trien System**

**Cases Where the Equilibrium
Concentration of the Ligand is not Known**

Whenever the total ligand concentration is about equal to or lower than that needed to form the fully coordinated metal complex, i.e., $C_L \leq n C_M$, the equilibrium concentration of ligand reflects the degree of complex formation and dissociation. This situation can be convenient handled by a **Ligand Balance Equation** (LBE). Analogous to the PBE (Chapter 4) the LBE matches those species which result from release ligand with those requiring additional ligand. The following examples will illustrate the principle.

## Example 5.4

We wish to find the composition of a solution that is 0.005 M $Cu^{2+}$ and 0.010 M in $NH_3$. The calculation proceeds by first calculating the $[NH_3]$, then using this value to obtain $\alpha$ values. We can start as if the solution consisted of 0.01 M $Cu(NH_3)_2^{2+}$. The LBE for this is

$$2[Cu(NH_3)_4^{2+}] + [Cu(NH_3)_3^{2+}] + [NH_3] = [Cu(NH_3)^{2+}] + 2[Cu^{2+}]$$
$$(5\text{-}5)$$

Starting with $Cu(NH_3)_2^{2+}$, production of the species on the left required addition of $NH_3$, those on the right resulted when $NH_3$ was released. Naturally, $[NH_3]$ itself is on the left; it requires $NH_3$.

Every term in Equation 5-5 can be expressed in terms of $[NH_3]$, the only variable in the $\alpha$ expressions. The solution to this problem can be found simply by using the pointer function, P, which here is

$$P = [Cu(NH_3)^{2+}] + 2[Cu^{2+}] - [Cu(NH_3)_3^{2+}] - 2[Cu(NH_3)_4^{2+}] - [NH_3]$$

A plot of @LOG(@ABS(P)) vs. log $[NH_3]$ as seen in Figure 5.1 shows that in 0.005 M $Cu(NH_3)_2^{2+}$ the $[NH_3]$ is $10^{-2.34}$. Figure 5.1 also allows us to see the concentrations of all the Cu-bearing species at this ligand concentration.

## Example 5.5

What is the pointer function for 0.005 M $Cu(NH_3)_3^{2+}$? The LBE and associated pointer function applicable in 0.005 M $Cu^{2+}$ containing a threefold excess of $NH_3$, which is equivalent to $Cu(NH_3)_3^{2+}$, can be seen as

$$[Cu(NH_3)_4^{2+}] - [NH_3] = [Cu(NH_3)_2^{2+}] + 2[Cu(NH_3)^{2+}] + 3[Cu^{2+}]$$

and

$$P = [Cu(NH_3)_2^{2+}] + 2[Cu(NH_3)^{2+}] + 3[Cu^{2+}] - [Cu(NH_3)_4^{2+}] - [NH_3]$$
$$(5\text{-}6)$$

## EFFECT OF pH ON METAL-COMPLEX EQUILIBRIA

Metal complexing agents (Lewis bases) are also Brønsted bases in many cases and will, therefore, be affected by changes in pH. For example, $NH_3$, $CN^-$, en, trien, and the EDTA anion will accept protons and hence the fraction of their total concentrations, which are available for metal complexation, vary with the pH.

This will pose no particular difficulty; we will simply employ the strategy described in Chapter 4. The concentration of the complexing species, $[L]$, is the product of its analytical concentration, $C_L$, and the fraction, $\alpha_L$, corresponding to the ratio of $[L]$ to $C_L$.

Let us illustrate this approach on the $Cu^{2+}$ - $NH_3$ system. In the example above, it was assumed that the pH was sufficiently high so that $\alpha_{NH3} = 1$. In general, however, we must write: $[NH_3] = \alpha_{NH3} C_{NH3}$. This is illustrated below.

### Example 5.6

Describe the solution composition for 0.005 M $Cu^{2+}$, 0.010 M $NH_3$ with the solution pH as 9.0. We again may recognize that this solution can be described as 0.005 M $Cu(NH_3)_2^{2+}$ so that the Pointer Equation 5-5 applies. The composition differs in that $[NH_3]$ for each Cu-containing species is replaced by 0.010 $\alpha_{NH3}$, where

$$\alpha_{NH_3} - \frac{Ka}{H+Ka} - \frac{10^{-9.24}}{10^{-9.0} + 10^{-9.24}} - 0.37$$

$$\text{Hence} \quad K_{f_1} - \frac{[Cu(NH_3)^{2+}]}{[Cu^{2+}][NH_3]} - \frac{[Cu(NH_3)^{2+}]}{[Cu^{2+}]\alpha_{NH_3}C_{NH_3}}$$

$$\text{or} \quad K_{f_1} - K_{f_1}\alpha_{NH_3} - \frac{[Cu(NH_3)^{2+}]}{[Cu^{2+}]C_{NH_3}}$$

$$\text{Similarly,} \quad K_{f_2} - K_{f_2}\alpha_{NH_3} - \frac{[Cu(NH_3)_2^{2+}]}{[Cu(NH_3)^{2+}]C_{NH_3}}$$

The conditional combined formation constants would then be

$$\beta'_i = K'_{f1}K'_{f2} \cdots K'_{fi} = \beta_i \, \alpha_{NH3}{}^i \qquad (5\text{-}7)$$

For example,

$$\beta'_4 = K'_{f1}K'_{f2}K'_{f3}K_{f4}' = \beta_4 \, \alpha_{NH3}{}^4$$

Notice that since each stepwise constant is altered by the $\alpha_{NH3}$ factor, the entire set of curves in Figure 5.1 would move horizontally (without changing their positions relative to each other) by log $\alpha_{NH3}$. This amounts to changing the quantity plotted on the X-axis from $\log[NH_3]$ to $\log C_{NH3}$.

## Example 5.7a

What are the concentrations of all species in a 0.005 M Cuen$^{2+}$ that is buffered at pH 9.0? (pK(en) = 7.52, 10.65.) The LBE, as before, is

$$[Cuen_2{}^{2+}] + [en] = [Cu^{2+}]$$

Hence, the curves in the Cu-en system (Figure 5-2) are shifted to the right by 1.65 units (where now the X-axis is in terms of

$$\alpha_{en} = \frac{K_{a_1}K_{a_2}}{H^2 + K_{a_1}H + K_{a_1}K_{a_2}} \cong 10^{-1.65}$$

Log $C_{en}$, not Log [en]. Nevertheless, since [en] is so small, the LBE can be simplified to

$$[Cuen_2{}^{2+}] \approx [Cu2+]$$

or

$$\alpha_{Cuen2}C_{Cu} = \alpha_{Cu}C_{Cu}$$

and, since $\alpha$s for both $[Cuen_2{}^{2+}]$ and $[Cu^{2+}]$ have the same denominator:

$$\beta_2\alpha^2_{en}C^2_{en} = 1$$

$$C_{en} = 1/\alpha_{en}\beta_2{}^{\frac{1}{2}}$$

Rearranging,

$$\alpha_{en} C_{en} = [en] = 1/\beta_2^{\frac{1}{2}}$$

Now here is a surprise! Is this the same answer as would be obtained if $\alpha_{en}$ were one? Certainly! **The concentration of [en] does not change with pH** (unless there is enough acid to almost totally reverse complex formation). The presence of the $Cu^{2+}$ would seem to hold [en] independent of pH changes.

## Example 5.7b

What about $0.005$ $Cuen_2^{2+}$ at pH $= 9.0$?

$$\text{LBE:} \quad [en] = [Cuen^{2+}] + 2[Cu^{2+}]$$

Here $[Cu^{2+}]$ is small enough so that

$$[en] - [Cuen^{2+}] - \alpha_{Cuen} C_{Cu} = \frac{0.005\, \beta_1 \alpha_{en} C_{en}}{1 + \beta_1 \alpha_{en} C_{en} + \beta_2 \alpha_{en}^2 C_{en}^2}$$

To a first approximation (see Figure 5.2)

$$[en] \text{ or } \alpha_{en} C_{en} = 1/K_{f2} \alpha_{en} C_{en}$$

Again, $\quad\quad\quad\quad\quad C_{en} = 1/\alpha_{en}\sqrt{K_{f2}}$

or $\quad\quad\quad\quad\quad [en] = \alpha_{en} C_{en} = 1/\sqrt{K_{f2}}$

  These results which show that a ligand that is also a Brønsted base will not have its concentration lowered by protonation in the presence of its metal complexes. The metal-ligand complex thus acts as a "ligand buffer" because the metal ion competes successfully with the proton for the ligand. This condition requires that the metal complex formation constant(s) as well as the metal ion concentration, $C_M$, be sufficiently high for the solution pH (as acidity increases, this condition becomes harder to maintain). We will encounter an analogous "metal buffering" when we deal with a solution containing more than one metal complexing agent.

# METAL COMPLEX EQUILIBRIA INVOLVING SEVERAL COMPLEXING AGENTS

Solutions containing more than one ligand are fairly common. For example, ground waters can have humic acids and ammonia, industrial effluents may have chlorides and cyanides, etc. In the course of an analytical reaction, ligands might be added as masking agents so that color-forming complexes will be limited to the metal ion understudy, or that a complex-forming titrant can be used for the metal of interest without interference.

In general, the values of $\alpha_M$ in mixtures of ligands can be obtained from the combination of $\alpha_M$ values for the individual ligands. After all, $\alpha_M$ is

$$\alpha_M = \frac{[M]}{[M] + [ML] + \cdots [ML_M] + [MX] \cdots [MX_M] + [MY] + \cdots}$$

Inverting we have

$$\frac{1}{\alpha_M} = \frac{[M] + [ML] + \cdots [ML_M] + [MX] \cdots [MX_M] + [MY] + \cdots}{[M]}$$

(5-8)

For each ligand

$$\frac{1}{\alpha_{M(L)}} = \frac{[M] + [ML] + \cdots [ML_N]}{[M]}$$

Examining Equation 5-8, we notice that the nominator has only one [M] term rather than N such terms needed when N ligands are present. This may be rectified by adding (N-1) [M] terms and subtracting an equal number. Then Equation 5-8 becomes

$$\frac{1}{\alpha_M} = \sum_i^N \frac{1}{\alpha_{M(L)_i}} - (N-1)$$

(5-9)

The second term in Equation 5-9 may be neglected in most problems, since when complexation is significant $1/\alpha_M > (N-1)$. In practice, one ligand is usually stronger than the rest so that $\alpha_M$ calculated for this ligand alone remains essentially the same after the other ligands are considered. This can easily be inferred from Equation 5-9.

## Example 5.8

Calculate the $\alpha_{Zn}$ for a solution in which the equilibrium concentration of EDTA, $[Y^+] = 10^{-4}$. Recalculate $\alpha_{Zn}$ when ammonia is also present at a level at which $[NH_3] = 10^{-2}M$

$$\alpha_{Zn(Y)} = \frac{1}{1 + \beta_{ZnY}[Y]^4} \approx \frac{1}{10^{16.5} \cdot 10^{-4}} = 10^{-12.5}$$

In the presence of both EDTA and $NH_3$,

$$\frac{1}{\alpha_{Zn}} = \frac{1}{\alpha_{Zn(Y)}} + \frac{1}{\alpha_{Zn(NH_3)}} - 1$$

For

$$\alpha_{Zn(NH_3)} = \frac{1}{1 + 10^{2.18}\ 10^{-2} + 10^{4.43}\ 10^{-4} + 10^{6.74}\ 10^{-6} + 10^{8.70}\ 10^{-8}}$$

$$= \frac{1}{1 + 1.51 + 2.69 + 5.50 + 5.01} = \frac{1}{15.7} = 10^{-1.2}$$

Therefore,

$$\frac{1}{\alpha_{Zn}} = \frac{1}{10^{-12.5}} + \frac{1}{10^{-1.2}} = 10^{12.5}$$

While the evidence for the constancy of $\alpha_{Zn}$ in the presence of excess EDTA regardless of the presence or absence of ammonia is clear and incontrovertible from this mathematical analysis, it may not be easy for you to accept. In a solution of $Zn^{2+}$, $ZnY^{2-}$, and excess EDTA, the calculation convincingly shows that the addition of $NH_3$ will not change $[Zn^{2+}]$ **at all**. Yet, the extent of complexation of $Zn^{2+}$ by $NH_3$ proceeds as usual. This must mean that the Zn for ammonia complexation must have come from $ZnY^{2-}$ ! That Zn leaves EDTA, the much stronger ligand, to join $NH_3$, the much weaker ligand, is a big surprise!

This "mystery" can readily be solved, however, by returning to our first principles, namely, the role of concentrations on the position of equilibrium. There is much more $ZnY^{2-}$ than $Zn^{2+}$ in the solution. Further, only a tiny fraction of $ZnY^{2-}$ is used to form the $Zn^{2+}$-ammonia complexes.

## CONDITIONAL FORMATION CONSTANTS

We have seen how the pH and the presence of several complexing agents can be taken into account in equilibrium calculations. If, as happens often, we can identify one reaction in the array of reactions as the principal reaction, then all the others can be properly be called side reactions, and treated in a convenient manner. For example, in complexometric titrations of metal ions with EDTA or some other polydentate chelating titrant, the presence of auxiliary ligands like $NH_3$, citrate anion, etc., can best be accounted for by the use of the **conditional constant**, first introduced by Schwarzenbach and widely applied by Ringbom.

By adopting this custom, we are essentially dividing reactions into two categories: (1) the main reaction, which occurs between the metal ion and ligand of interest and (2) side reactions. Side reactions include all other possible reactions of (a) the metal ion, such as with OH⁻ or other ligands, (b) the ligand, such as with $H^+$ or other metal ions, and even of (c) the metal complex, which in some cases form additional products with $H^+$, OH⁻ Cl⁻, etc.

Let us illustrate this by the formation equilibrium for ML from M and L using the equivalent $\alpha \cdot C$ expressions.

$$\beta_1 = \frac{[ML]}{[M][L]} = \frac{\alpha_{ML} C_{ML}}{\alpha_M C_M \cdot \alpha_L C_L}$$

$$\beta_1' = \beta_1 \frac{\alpha_M \alpha_L}{\alpha_{ML}} = \frac{C_{ML}}{C_M C_L} \tag{5-10}$$

The symbol $\beta'$ is called the **conditional** formation constant and is different from the formation constant in the presence of side reactions, since the $\alpha$ values will be different from unity and would depend on the reaction **conditions** such as pH, etc. Provided the reaction conditions remain constant in the problem under study, $\beta'$ remains constant. Why do we bother with such a changeable "constant"? For one thing, it greatly simplifies the stoichiometry. For example, in a solution containing 0.001M $NiY^{2-}$ (where Y is the EDTA anion) at pH 9.0 containing 0.02 M $NH_3$, the dissociation of $NiY^{2-}$, yields a number of different Ni-containing species (the various $Ni(NH_3)^{2+}$; complexes) as well as various protonated EDTA species. Therefore, it would be grossly incorrect to say $[Ni^{2+}]$ equals $[Y^{4-}]$ but with the conditioned constant formulation, we can say $CY = C_{Ni}$. All of the species are conveniently taken care of by the $\alpha C_2$ notation; that is, $[Ni^{2+}] = \alpha_{Ni}C_{Ni}$ and $[Y^{4-}] = \alpha_Y C_Y$.

We must be alert to the following changes, however. While $\alpha$ values are calculated as before, the term, $C_L$, takes on new meaning. C no longer represents the **total** analytical concentration. For example, none of the ligand species which are bonded to the metal ion in question are included. For instance, in the examples above,

$$C_{NH3} = [NH_3] + [NH_4^+]$$

**does not** include any of the $[Cu(NH_3)_i^{2+}]$, and

$$C_Y = [Y^{4-}] + [HY^{3-}] + \ldots [H_4Y]$$

**does not** include $[CuY^{2-}]$.

Similarly, while $C_{Ni}$ includes both $[Ni^{2+}]$ and all the "side reaction" complexes, it does not include $[NiY^{2-}]$.

## Example 5.9

In a solution containing 0.010 M $Ni(NO_3)_2$, 0.015 M $Na_2H_2Y$, 0.10 M $NH_3$ with sufficient $NH_4Cl$ to make the pH 9.0, calculate $[Ni^{2+}]$.

First, we must recognize the major composition change resulting from the formation of $NiY^{2-}$, the principal reaction makes $C_Y$ become 0.005 M, the remainder having reacted with $Ni^{2+}$. The expression for the conditional constant, $\beta'_{NiY}$, is

$$\beta_{NiY} = \beta_{NiY} \frac{\alpha_{Ni} \, \alpha_{Y}}{\alpha_{NiY}} = \frac{C_{NiY}}{C_{Ni} \cdot C_{Y}}$$

where $C_{NiY} = 0.010$, $C_Y = 0.005$ and $C_{Ni}$ is to be evaluated as a step in finding $[Ni^{2+}]$. The value of $\beta'_{NiY}$ requires the calculation of $\alpha_Y$ from the pH and $\alpha_{Ni}$ from $[NH_3]$ in the usual fashion; since $NiY^{2-}$ does not enter into any side reactions, $\alpha_{NiY} = 1$.

Using pK (EDTA) values of 2.0, 2l7, 6.13, and 10.33, we can see merely by drawing the system line (as under EquiligrapH construction, see Chapter 4) and noting that $HY^{3-}$ not only predominates at pH = 9.0, but exceeds any other EDTA species by at least 20-fold (1/20 = 5%), so that $\alpha_Y \cong K_1K_2K_3K_4/K_1K_2K_3H = K_4/H = 10^{-1.33}$.

Since $[NH_3] = 0.10$ (enough $NH_4+$ to prevent significant further transformation of $NH_3$ to $NH_4^+$), $\alpha_{Ni}$ can be readily evaluated.

$$\alpha_{Ni} = 1/(1 + \beta_1[NH_3] + \beta_2[NH_3]^2 + \ldots + \beta_6[NH_3]^6)$$

where $\log\beta_1$ from i = 1 to 6 are 2.72, 4.89, 6.55, 7.67, 8.34, 8.31 and $[NH_3] = 0.1$, then $\alpha_{Ni} = 10^{-4.07}$.

These $\alpha$ values with the value of $\beta_{NiY} = 10^{18.62}$, gives a value of $\beta'_{NiY} = 10^{18.62} \cdot 10^{-1.33} \cdot 10^-$. By rearrangement, $C_{Ni} = C_{NiY}/C_Y \cdot \beta'_{NiY} = 0.010/0.0005 \cdot 10^{+13.22} = 10^{-12.92}$

$$[Ni^{2+}] = \alpha_{Ni}C_{Ni} = 10^{-4.07} \cdot 10^{-12.92} = 10^{-16.99}$$

Notice that since $\beta'_{NiY}$ has $\alpha_{Ni}$ as a factor, $C_{Ni}$ is proportional to $1/\alpha_{Ni}$. Hence, since $[Ni^{2+}]$ equals $\alpha_{Ni}C_{Ni}$, in the presence of excess EDTA, the value of $[Ni^{2+}]$ is independent of the value of $\alpha_{Ni}$, that is, of the presence or absence of $NH_3$. We may draw a general conclusion from this example, that the value pM($-\log[M_n]$) is a mixture of a metal ion with an excess of a strong complexing agent remains unchanged by the addition of other metal complexing agents provided that the others do not form as stable complexes as the first. Thus, the system may properly be called a "metal buffer". Naturally, the pH affects the conditional stability constant of the principle reaction, pM will change accordingly.

## PROBLEMS

**5-1**  Develop equations relating logß to log*ß as a function of ionic strength (as in Chapter 3 for FeY⁻) for the formation of the following complexes:

(a) $Cu(NH_3)_4^{2+}$            (d) $AgCl2^-$
(b) $FeCl4^-$                    (e) $NiY^{2-}$(EDTA ion)
(c) $Zn(CN)_4^{2-}$              (f) $Cu(8\text{-hydroxyquinolinate})_2$

**5-2**  Construct and solve problems like that in Example 5.1 with the following systems:

(a) Fe(III) - Cl⁻               (e) Ni(II) - $NH_3$
(b) Fe(III) - Br⁻               (f) Ag(I) - $NH_3$
(c) Cu(I) - Cl⁻                 (g) Pb(II) - Cl⁻
(d) Co(II) - Cl⁻                (h) Ni(II) - CN⁻

In these and similar problems, use Block Names(/ENC{/BAC; /RNC} or absolute cell addresses so that you may change concentrations and formation constant values (for the different parts of the problem) with minimum effort.

**5-3**  a)    Develop the EquiligrapH (log concentration of all metal species as a function of log ligand concentration) for solutions of 0.02 M $Cd(NO_3)_2$ as a function of equilibrium ammonia concentrations from -5 to +1.
       b)    By means of a ligand balance equation (LBE) and the pointer function, find pCd($-\log[Cd^{2+}]$) for a solution in which initial concentrations of $Cd^{2+}$ is 0.02 M and $NH_3$ is 0.06 M.
       c)    Repeat (b) if the pH is maintained at 8.8.

**5-4**  Draw, using the spreadsheet, plots of the $\alpha_{MLi}$ values in the systems described in Example 5.1. Compare these distribution diagrams with the equiligrapH shown in Figure 5.1.

**5-5**  Develop a spreadsheet and corresponding graph for a plot of the conditional formation constants, log $\beta'_{CuY}$ as a function of pH for EDTA, trien, tetren. What is the significance of the intersections on these graphs?

**5-6**    Draw with the help of a spreadsheet 2 graphs showing the effect(s) of pH on the log $\beta'$ values of 12 metal EDTA complexes.

**5-7**    Find the conditional constants for the following ligand transfer reactions as a function of pH.

$$ML + L' = ML' + L$$

Where M can be $Cu^2 + Zn^{2+}$, and L = EDTA or HEDTA and L' = trien or tren.  Compare the systems by noting the pH values at which the ligand transfer constant becomes one.

# Chapter 6

# Precipitation Equilibria

Precipitation formation and the solution of precipitates are probably the most familiar of all separation processes. Precipitation represents a simple means of isolating both individual ions and groups of ions, and has been utilized by analytical chemists for a wide range of substances. The classical beauty of the general "qual" scheme of undergraduate qualitative inorganic analysis has, in its time, convinced many students to follow a career in chemistry. It utilizes sequential precipitation by a small number of reagents to separate metal ions into groups. For example, addition of hydrochloric acid will result in the precipitation of chlorides, $MCl$, for the singly charged ions, $Cu^+$, $Ag^+$, $Au^+$, and $Tl^+$, as well as $Hg_2^{2+}$. When the solution is acidified with HCl, a number of additional compounds precipitate, the acid-insoluble hydroxides (also referred to as the oxyacids) of Si(IV), Ti(IV), V(V), Nb(V), Ta(V), and W(VI).

The second group of metal ions consists of those elements whose sulfides are so insoluble that $H_2S$ in 0.3 M HCl will precipitate them:

|    |    |    |    | Cu | Zn | Ge | As | Se |
|----|----|----|----|----|----|----|----|----|
| Mo | Tc | Ru | Pd | Ag | Cd | Sn | Sb | Te |
| W  | Re | Os | Pt | Au | Hg | Pb | Bi | Po |

Yet another group of metal ions is separated as hydroxides (i.e., hydrous oxides) when, after the $H_2S$ is removed and an oxidant is added, the solution is made alkaline with $NH_3$. $Fe(OH)_3$, $Mn(OH)_3$, $Al(OH)_3$, $Cr(OH)_3$, and $NH_4VO_3$ will precipitate. Adding hydrogen sulfide to the filtrate precipitates $CoS$, $NiS$, and $ZnS$. Carbonate addition to this filtrate will bring down the alkaline earth metal (Mg, Ca, Ba, Sr) carbonates, leaving the alkali metal ions (Li, Na, K, Cs) remaining in solution.

Once separated into these major groups, further separation is obtained by differentiating the sulfides and the hydroxides by their amphoteric properties, i.e., their solubility in NaOH, then followed by more selective precipitants. It is interesting that the solubilities of metal sulfides, hydroxides, and other salts vary with their electronegativities as well as those of the precipitating anions in a manner similar to that described for the strength of metal complexes (Chapter 5).

The "qual scheme" served as the basis for the development of radiochemical analysis by the chemists of the Manhattan Project in the 1940s. There are a number of industrial processes including those utilized in hydrometallurgy that are based on precipitation steps to produce the desired intermediate or product. In nature, production of various minerals from natural waters and brines can be understood in terms of precipitation equilibria (and kinetics).

Consideration of precipitation also plays a major role in describing and understanding selective solution of precipitates and minerals in the laboratory, process plant, and in the environment.

## Factors That Affect the Solubility of Electrolytes

The solubility of a solid in water depends upon the difference between the energy consumed in separating the ions or molecules from the crystal lattice, (i.e., the lattice energy), and the energy released by the solvation of these ions or molecules (called the hydration energy). The lattice energy of strong electrolytes increases with the ionic charge and decreases with the ionic size. Thus, $CaO$ has a higher lattice energy than $NaCl$ which in turn has a higher lattice energy than $KCl$. Similarly, hydration energies are largest for small, highly charged ions. Since hydration energies and lattice energies vary in the same manner, it is difficult to predict solubility trends in various types of electrolytes from such data.

With many electrolytes, the lattice energy is somewhat greater than the hydration energy so that the dissolution of electrolytes in water is generally an endothermic process. Hence, the solubility of most electrolytes increases with increasing temperature. See Equation 2-13. A very few substances such as $CaSO_4$, have negative heats of solution and therefore exhibit a decrease in solubility with increase in temperature.

The solubilities of solid electrolytes in organic solvents are generally lower than they are in water. Thus the solubility of $PbSO_4$ or $SrSO_4$ can be decreased by the addition of ethanol to water, which thereby facilitates their quantitative precipitation. The solubility decreased effected by the organic solvent probably reflects the influence of the lower dielectric constant which increased the energy required to dissociate the ions. Just what effect the change of solvent has on the solubility of molecules (or uncharged ion pairs) is far too complex to be considered here, and in any event is usually of much less importance.

Since solubility of electrolytes involves dissociation phenomena, factors affecting dissociation either directly (ionic strength) or indirectly (reactions involving ions formed) will inevitably affect solubility. These factors will now be considered in detail.

## Solubility and Solubility Product

The solubility of a substance at any given temperature is defined as the concentration of the substance in a saturated
solution, i.e., a solution which is in _equilibrium_ with the undissolved solid. This equilibrium may be represented as follows, using a solution of phenol, $(C_6H_5OH)$ in water as an illustration.

$$(C_6H_5OH)_{solid} \rightleftharpoons (C_6H_5OH)_{solution}$$

with the corresponding equilibrium expression

$$.^*K - a_{C_{12}H_{10}} - [C_{12}H_{10}] \; \gamma_{C_{10}H_8} \qquad (6\text{-}1)$$

The activity of the solid, being a constant, is included in the equilibrium constant, K. A distinction should be made between saturated and concentrated solutions since these categories are not equivalent. Here, for example, the low solubility of phenol in water results in a saturated solution of approximately $10^{-4}$ M, which is extremely dilute.[1]

If, as in the case of phenol, we are dealing with solutes for which the activity coefficients are very close to unity, then the equilibrium constant, $^*K$, can be recognized as the molar solubility. However, it should be kept in mind that the activity coefficients of even undissociated solutes will be affected to some extent by the total solution composition. Hence, solubilities of nonelectrolytes are not absolutely invariant at a constant temperature.

The foregoing discussion of solubility does not cover the case of the solubility of a strong electrolyte since strong electrolytes will dissociate completely in solution. The concept of total dissociation of strong electrolytes in aqueous media has been modified in recent years by the admission of the existence of ion pairs or other ion-association complexes. Such species, although differing from coordination complexes in the nature of

---

[1]Phenol is a weak monoprotic acid. How serious a mistake have we made in ignoring the effect of its dissociation on its solubility in water?

their bonding, are described by the same type of equilibrium expressions (see Chapter 5). NaCl in water for example may be represented by the following equations.

$$(NaCl)_{solid} \rightleftharpoons (NaCl)_{solution}$$

However, since $(NaCl)_{solution}$ is totally dissociated into $Na^+$ and $Cl^-$ ions, this equation must be rewritten as

$$(NaCl)_s \rightleftharpoons Na^+ + Cl^-$$

The equilibrium constant here is

$$.^*K_{sp} - a_{Na} \cdot a_{Cl^-} - m_{Na} m_{Cl} \ \gamma_{Na} \cdot \gamma_{Cl^-} \tag{6-2}$$

where m, molality, is used instead of molarity because the solubility is so high that the usual assumption that m and M are equal is not valid.

As before, the activity of the solid NaCl is a constant and is included in the equilibrium constant. In this case, in contrast with that of the nonelectrolyte, the equilibrium constant may not be equated to the solubility because the right-hand side of the equation contains the **product** of two concentration terms rather than a single concentration term. This equilibrium constant is called the solubility product constant $^*K_{sp}$. An added complexity arises in the illustration that we have chosen, because a saturated solution of NaCl happens to be quite concentrated. Hence the activity coefficients included in the equation may be quite different from those predicted by the Davies' Equation, rendering attempts to describe quantitatively the solubility relationships of NaCl with the use of this equation difficult. The solubility product equilibrium just described is much more useful when applied to systems of **slightly soluble electrolytes**.

With $BaSO_4$, for example, which has a solubility of $1.0 \times 10^{-5}$ mol/L in water, we may write, as we did for NaCl:

$$(BaSO_4)_s \rightleftharpoons Ba^{2+} + SO_4^{2-}$$

$$.^*K_{sp} - [Ba^{2+}][SO_4^{2-}]\gamma_{Ba^2} \cdot \gamma_{SO_4^{2-}} \ ; \ \therefore \ K_{sp} - \frac{.^*K_{sp}}{\gamma_{Ba^2} \cdot \gamma_{SO_4^{2-}}} \tag{6-3}$$

In the remainder of this chapter, concentration constants, $K_{sp}$, calculated from $^*K_{sp}$ by methods described in Chapter 3, will be employed. These will vary with ionic strength (see Equation 3-15).

## Factors That Affect the Solubility Product Constant

The true or thermodynamic solubility product constant $K_{sp}$ will, for most substances, increase with the temperature by an amount that depends on the heat of solution. The addition of any organic solvent such as alcohol to an aqueous solution will generally result in a lower $^*K_{sp}$. This may be easily understood in terms of the increased work of separation of ions in a medium of lower dielectric constant.

The particle size of the undissolved solid in equilibrium with the dissolved solute can also affect the solubility and solubility product constant. If the particle size is **sufficiently** small, the surface area per mole becomes large enough to require taking the surface energy into account in describing the equilibrium. In effect, the smaller the particle size (i.e., smaller than $10^{-4}$ cm radius), the higher the solubility. For example, the molar solubility of lead chromate having particles of $9 \times 10^{-5}$ cm radius was found to be $2.1 \times 10^{-4}$ in contrast to a value of $1.24 \times 10^{-4}$ for particles having a radius of $3.0 \times 10^{-3}$ cm or larger. In this case, the $K_{sp}$ has changed by 250%. There are probably cases in which the effect of particle size on $K_{sp}$ is much larger, but these are exceedingly difficult to verify experimentally.

Calculations involving precipitation equilibria involve yet another difficulty, namely the slow equilibration of the solid phase with the solution. Thus, kinetic effects may give rise to results which deviate significantly from those calculated.

In dealing with solubility equilibria, it is usually assumed that the solid phase or precipitate is pure. In practice this is rarely achieved and at best only approximately so. For this reason separations that are predicted without taking such contamination into account are never fully realized.

The effect of changing ionic strength on the value of the concentration solubility product constant $K_{sp}$, already briefly mentioned above, is even more significant than the corresponding effect on monoprotic acid dissociation constants. This is to be expected with precipitates containing multicharged ions. (See Equation 3-11 and forward.)

### Table 6.1 Dielectric Constant, $\epsilon$, and CaSO$_4$ Solubility

| C$_2$H$_5$OH | $\epsilon$ | Solubility |
|---|---|---|
| 0.0 w/w% | 80 | 2.084 g/L |
| 3.9 | 78 | 1.314 |
| 10.0 | 73 | 0.970 |
| 13.6 | 71 | 0.436 |

## Rules for Precipitation

The solubility product expression can be used for predicting whether or not precipitation will occur upon mixing two solutions and whether or not a precipitate will dissolve when in contact with a given solution. This represents an application of the general criteria for predicting the direction of reactions and was developed earlier. For the purposes of this discussion, it will be convenient to call the product of the concentrations of the ions each raised to the appropriate power, i.e., the right-hand side of the solubility product expression,

$$K_{sp} = [M]^a [X]^b, \text{ which is the } \underline{\text{ion product}}.$$

The ion product would be the actual value (not the equilibrium value), assuming that no precipitation occurred. The following relations will then be seen to apply.

Ion product < $K_{sp}$:     **Solution is unsaturated.**
No precipitate will form and precipitate present will dissolve.

Ion product > $K_{sp}$:     **Solution is supersaturated.**
In time a precipitate will form, and precipitate present will not dissolve.

Ion product = $K_{sp}$:     **Solution is saturated.**
In this equilibrium mixture, no precipitate will form and precipitate present will not dissolve.

## Quantitative Relationship Between Solubility and Solubility Product

It may readily be shown that the $K_{sp}$ is not the solubility, but it is, of course, related to it. In a saturated solution of $BaSO_4$ of $1.0 \times 10^{-5}$, molarity, the concentration of $Ba^{++}$ and of $SO_4^=$ is each $1.0 \times 10^{-5}$, since $BaSO_4$ is completely dissociated. Hence,

$$K_{sp} = 1.0 \times 10^{-5} \times 1.0 \times 10^{-5} = 1.0 \times 10^{-10}$$

We can generalize this for any slightly soluble strong electrolyte MX whose molar solubility is S, that $K_{sp} = [M][X] = S^2$.

In the most general case where $M_aX_b$ is the formula of a slightly soluble strong electrolyte, with a solubility S mol/L,

$$(M_aX_b)_{solid} \rightleftharpoons aM + bX$$

for every mole of $M_aX_b$ dissolved $a$ moles of M and $b$ moles of X are formed. Hence the solubility product expression,

$$K_{sp} = [M]^a[X]^b$$

$$= (aS)^a \cdot (bS)^b$$

$$= a^a \cdot b^b \cdot S^{(a+b)} \tag{6-4}$$

It is important to recognize that although we have multiplied S by the coefficient $a$ and then raised the product to the power of $a$, that this is **not** equivalent to saying that the concentration of M was multiplied by $a$ before being raised to the power of $a$. The concentration of M is (aS), **not** S. For example, when a salt such as $M(NO_3)_n$ is present with the precipitate, the *total* $C_M$ is $(C_M + aS)$.

The equations relating the molar solubility S in water to the $K_{sp}$ of the following compounds are

$$Ag_2CrO_4: \quad K_{sp} = [Ag^+]^2[CrO_4^=] = (2S)^2(S) = 4S^3$$

$$Ce_2(C_2O_4)_3: = K_{sp} = [Ce^{+3}]^2[C_2O_4^=]^3 = (2S)^2(3S)^3 = 108S^5$$

$$Hg_2Cl_2: K_{sp} = [Hg_2^{2+}][Cl^-]^2 = (S)(2S)^2 = 4S^3$$

$$MgNH_4PO_4: K_{sp} = [Mg^{2+}][NH_4^+][PO_4^{3-}] = S^3$$

Note that in the various cases the $K_{sp}$ is equated to the solubility raised to different powers, depending on the charge type of the precipitate. Hence, once cannot directly compare solubilities of salts of different charge types by examination of the numerical values of the respective solubility product constants.

How do each of these $K_{sp}$ values vary with ionic strengths (plot log $K_{sp}$ vs. I up to 0.5 using the Davies' Equation)?

These expressions correctly describe the relationship between solubility product and solubility either in pure water or in solutions that do **not** contain any other source of the ions directly involved in the solubility equilibrium. These expressions must be modified when salts containing common ions are present in solution.

## The Common Ion Effect

The common ion effect may best be illustrated by considering the solubility of $M_aX_b$ in the presence of $C_x$ mol/L of anion X. Then,

$$K_{sp} = (aS)^a \cdot (C_x + bS)^b \qquad (6-5)$$

In most cases that will be encountered, $C_x >> bS$, so that this equation simplifies to

$$K_{sp} = (aS)^a \cdot (C_x)^b \qquad (6-6)$$

## Example 6.1

Calculate the molar solubilities, S, of AgCl and $Ag_2CrO_4$ under the following conditions.

(a)     Compare the solubility of AgCl in 0.01 M $NaNO_3$ with those in either 0.01 M $AgNO_3$ or 0.01 M NaCl. In $NaNO_3$, in the absence of an electrolyte having a common ion, S is obtained from $K_{sp} = 10^{-9.92} = S^2$, so that $S = 1.1 \times 10^{-5}$. In either of the other two electrolytes, however, the proper expression is $K_{sp} = 10^{-9.92} = S(S+0.01) \approx 0.01S$.

(b)     Compare the solubility of $Ag_2CrO_4$ in 0.01 M $NaNO_3$ with those in either 0.01 M $AgNO_3$ or 0.01 M $Na_2CrO_4$. In the electrolyte without a common ion, the expression is $K_{sp} = 10^{-11.35} = 4S^3$, and $S = 1.1 \times 10^{-4}$. In 0.01 M $AgNO_3$, $K_{sp} = (0.01 + 2S)^2(S) \approx (0.01)^2S$, and $S = 4.5 \times 10^{-8}$. The solubility of $Ag_2CrO_4$ in 0.01

M $Na_2CrO_4$ is given by $K_{sp} = (2S)^2(0.01+S) \approx 4S^2 \times 0.01$. Hence, $S = 1.4 \times 10^{-5}$. (Please note that the $K_{sp}$ used here, was larger than in the earlier parts of this question, because of the larger ionic strength.)

The common ion effect may be considered to apply to any solution in which there is more than one source of the ions contained in the precipitate. The common ion effect is involved in such questions as, what concentration of reagent anion would be necessary to initiate precipitation, and what is the anion concentration when the precipitation is (virtually) complete?

## Example 6.2

(a)    What concentration of $CO_3^{2-}$ is necessary to initiate precipitation of $CaCO_3$ from a 0.01 M $Ca(NO_3)_2$ solution? $pK_{sp}$ for $CaCO_3$ in this solution is 7.73.

(b)    What is the concentration of $CO_3^{2-}$ when the $Ca^{2+}$ has been quantitatively precipitated (i.e., only 0.1% $Ca^{2+}$ remains in solution)?

(c)    When the solution is saturated with $CaCO_3$, we may write $[Ca^{2+}][CO_3^{2-}] = 10^{-7.73}$.

Since $[Ca^{2+}] = 0.01$, $[CO_3^{2-}]$ necessary for saturation $= 10^{-7.73}/0.01$, i.e.,

$$[CO_3^{2-}] = 10^{-5.73} = 1.9 \cdot 10^{-6} \text{ M}$$

At this concentration, the solution is saturated; no precipitate will form. When the concentration is slightly higher than this, precipitation will be initiated.

(d)    In order to describe the conditions for quantitative precipitation, we must first agree on a definition of quantitative precipitation. You know we cannot define this as remaining $[Ca^{2+}] = 0$, since the $K_{sp}$ equation predicts that this would require an infinite concentration of $CO_3^{2-}$. Several practical definitions of quantitative precipitation have been widely used. (1) All but a small fraction of the original concentration, e.g., 0.1% or 0.01%, of the ion has been precipitated, (2) all but a given weight, e.g., 0.1 mg, of the ion has been precipitated. The definition chosen depends on its suitability to the problem at hand.

When the $[Ca^{2+}]$ in solution has dropped to 0.1% of the original 0.01 M, namely

$$[Ca^{2+}] = 0.001 \times 0.01 = 1.0 \times 10^{-5}$$

Then

$$[CO_3]^{2-} = 10^{-7.73}/1.0 \cdot 10^{-5} = 10^{-2.73} = 1.9 \cdot 10^{-3} \text{ M}$$

It has been assumed that the same value of $pK_{sp}$ applied in this solution also.

## Selective Precipitation

Having considered the precipitation of single substances, we are now ready to examine selective precipitation procedures for separating ions, a subject of even greater interest to the analytical chemist.

Consider, for example, a mixture of cations that can form precipitates with a particular reagent anion. Let us examine the changes that take place in this solution as we add this reagent in small increments. The first significant change occurs when the reagent concentration reaches a value sufficiently high for an ion product of one of the substances to just exceed its solubility product constant. This substance will precipitate first. The substance having the smallest solubility product constant will **not always** be the first to precipitate. What is always true, however, is that the substance requiring the least amount of reagent to reach saturation will be the first to precipitate.

## Example 6.3

Chloride ion is added to a mixture containing 0.01 M $Tl^+$, 0.02 M $Pb^{2+}$ and 0.03 M $Ag^+$. Calculate the order in which precipitation of these metal halides occur and to what extent separations of the three metals take place. $pK_{sp}$ of $TlCl$ = 3.46; $pK_{sp}$ for $PbCl_2$ = 4.08; $pK_{sp}$ for $AgCl$ = 9.50.

Using the solubility product expressions for each of the metal chlorides we can calculate the [Cl⁻] necessary for saturating the solution with respect to each salt.

$$[Tl^+][Cl^-] = 10^{-3.46}$$

$$\therefore \ [Cl^-] = 10^{-3.46}/0.01 = 10^{-1.46}$$

$$[Pb^{2+}][Cl^-]^2 = 10^{-4.08}$$

$$\therefore \ [Cl^-] = 10^{-1.19}$$

$$[Ag^+][Cl^-] = 10^{-9.50}$$

$$\therefore \ [Cl^-] = 10^{-7.98}$$

Hence, the order in which the metal chlorides will precipitate is AgCl, followed by TlCl, with PbCl₂ the last to precipitate, in keeping with the increasing concentration of [Cl⁻] required for saturation. As you see, this is not the order of increasing solubility product constants! (Why not? Is there more than one reason?)

The selectivity of this precipitation is described by how much of each metal ion will have precipitated before the next one begins to precipitate.

As soon as the [Cl⁻] has exceeded $10^{-7.98}$ mol/L, AgCl will begin to precipitate out and thereby lower [Ag⁺] in solution. This in turn will result in the requirement of a higher [Cl⁻] in order to continue the precipitation. Despite the fact that [Cl⁻] will increase, no TlCl will accompany the AgCl until [Cl⁻] reaches a value of $10^{-1.46}$. At this point, the [Ag⁺] left in solution must be

$$[Ag^+] = 10^{-9.50}/10^{-1.46} = 10^{-8.04} = 9 \cdot 10^{-9} \ M$$

This represents 100 $(9 \times 10^{-9}/0.03) = 3 \times 10^{-5}\%$ of the original amount present. This easily meets any reasonable criterion for quantitative removal of Ag⁺ from the solution before either Tl⁺ or Pb⁺⁺ begin to precipitate.

Similarly, we can calculate that the [Tl⁺] remaining in solution when PbCl₂ begins to precipitate is

$$[Tl^+] = 10^{-3.46}/10^{-1.19} = 10^{-2.27} = 5.4 \cdot 10^{-3} \ M$$

This represents a $5.4 \cdot 10^{-3}/0.01 \cdot 100$ or 54% Tl⁺ that remains unprecipitated at this point; although some separation does take place, not at all satisfactory for quantitative separation of $Tl^+$ and $Pb^{2+}$.

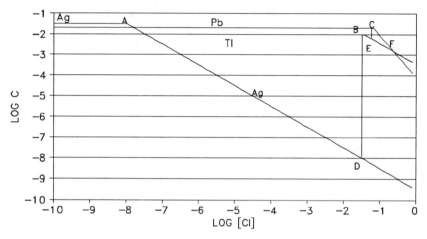

**Figure 6.1 Solubilities of Metal Chlorides**

A graphical representation of the solubility product relationships for the three chlorides can be used as the basis of the solution of the problem just considered. In Figure 6.1, horizontal lines are shown which terminate in lines of slope, - 1, - 1, and - 2, at points A, B, and C respectively. These lines of negative slope are obtained by plotting the logarithmic expressions of the corresponding solubility product expressions.

$$\log [Ag^+] \quad = -9.50 - \log [Cl^-]$$

$$\log [Tl^+] \quad = 3.46 - \log [Cl^-]$$

$$\log [Pb^{2+}] = -4.08 - 2 \log [Cl^-]$$

It is of interest to note that the difference in the nature of the dependence of the solubility upon the $[Cl^-]$ results in lines of different slope.

The significance of points A, B, and C is that they represent points at which the solution is saturated with respect to each of the salts and therefore show the $[Cl^-]$ at which precipitation for each is initiated. The lines beyond these points demonstrate how the metal ion concentration decreases with increasing $[Cl^-]$.

The question as to which of the three metals will precipitate first is solved simply by comparing the points of departure from the horizontal, A, B, and C. Obviously, the point occurring at the smallest [Cl⁻] will be that of the metal that precipitates first, in this case $Ag^+$.

Figure 6.1 also shows that $Tl^+$ is the next to precipitate. At this point B, $[Ag^+]$ may be readily obtained by dropping the perpendicular to point D as shown. Since the difference in log $[Ag^+]$ between point A (initial concentration) and point D (concentration at which TlCl begins to precipitate) is greater than 3 units (corresponding to 0.001 as the fraction of $Ag^+$ remaining in solution) it may be concluded that a quantitative separation of $Ag^+$ and $Tl^+$ is theoretically possible. Proceeding in a similar manner, the $[Tl^+]$ at which $PbCl_2$ begins to precipitate shown at point E is sufficiently close to the initial concentration (at point B) to predict that the quantitative separation of $Tl^+$ and $Pb^{++}$ as chlorides is not feasible.

The difference in the slopes of the TlCl and $PbCl_2$ lines causes the extent of the separation to depend on the initial metal concentration. Assuming that both $Pb^{++}$ and $Tl^+$ are present in the same initial concentration, TlCl can be seen to precipitate first, so long as this concentration is higher than its value at point F where the two lines intersect. At initial concentrations below this point $Pb^{++}$ will precipitate first.

Another approach to predicting the order of precipitation involves the use of the condition of simultaneous precipitation as the reference point. As an illustration the question of the order of the precipitation of silver salts from a solution that is 0.01 M in Cl⁻ and Br⁻, each, will now be considered. When the solution is saturated with respect to both salts, simultaneously precipitation of AgCl and AgBr will occur. At this point both $K_{sp}$ expressions apply.

$$[Ag^+][Cl^-] = K_{sp_{AgCl}} \quad and \quad [Ag^+][Br^-] = K_{sp_{AgBr}} \qquad (6\text{-}7)$$

Dividing one equation by the other, notice that the $[Ag^+]$ cancels, resulting in the following expression:

$$\frac{[Cl^-]}{[Br^-]} = \frac{K_{spAgCl}}{K_{spAgBr}} \qquad (6\text{-}8)$$

From this equation we can develop rules of fractional precipitation, analogous to the rules of precipitation developed at the beginning of this chapter. Let us call the ratio of the concentrations of the two ions the ion ratio, (in analogy to the term in product used earlier). If the ion ratio is equal to the solubility product ratio, then simultaneous precipitation will occur. If the ion ration is greater than the solubility product ratio, then precipitation of the salt whose ion concentration is in the numerator will occur first. Conversely, if the ion ratio is smaller than the solubility product ratio, the salt whose ion concentration is in the denominator will precipitate first.

For pairs of salts of the same charge type, the ion ratio required for simultaneous precipitation is independent of the total concentration, whereas for a pair involving different charge types the ratio depends on concentration. Thus for a mixture of $CrO_4^=$ and $Cl^-$ to which $Ag^+$ is added:

$$[Ag^+][Cl^-] - K_{spAgCl} \quad and \quad [Ag^+]^2[CrO_4^-] - K_{spAg_2CrO_4} \qquad (6\text{-}9)$$

The $[Ag^+]$ can be eliminated between these two equations to give:

$$\frac{[Cl^-]^2}{[CrO_4^-]} - \frac{(K_{spAgCl})^2}{K_{spAg_2CrO_4}} \qquad (6\text{-}10)$$

As before, comparison of the function $[Cl^-]^2/[CrO_4^=]$ with the value of the corresponding solubility product ratios may be used in predicting precipitation orders.

## Example 6.4

In a mixture containing chloride and bromide ions in a molar ration of 1000:1, what will precipitate first upon the addition of silver ions? $pK_{sp}$ of AgCl = 9.52 and $pK_{sp}$ of AgBr = 12.04. (These constants apply to a solution whose ionic strength has to be assumed to be 0.1.)

$$[Ag^+][Cl^-] = 10^{-9.52}$$

$$[Ag^+][Br^-] = 10^{-12.04}$$

Therefore, for simultaneous precipitation of AgCl and AgBr,

$$[Br^-]/[Cl^-] = 10^{-12.04}/10^{-9.52} = 10^{-2.52}$$

Since the actual ratio present in solution is $10^{-3}$ the addition of $Ag^+$ will produce a precipitate of the salt whose anionic concentration appeared in the denominator, namely AgCl.

The concentration of $Cl^-$ will decrease through the precipitation of AgCl alone, while the actual ratio reaches the value $10^{-2.52}$. In this case, the decrease in the $[Cl^-]$ is equivalent to $10^{-2.52}/10^{-3} = 10^{0.48} = 3.0$.

Hence, in this case two thirds of the chloride will precipitate before simultaneous precipitation of AgCl and AgBr occurs.

The following example will illustrate the influence of total concentration on the order of precipitation of salts of different charge types.

## Example 6.5

In a mixture containing $CrO_4^{2-}$ and $Cl^-$ in a molar ratio of 1000:1, which salt will precipitate first on the addition of $Ag^+$, if the initial $[CrO_4^{2-}]$ is (a) $1.0 \times 10^{-4}$ M and (b) $1.0 \times 10^{-2}$ M. $pK_{sp}$ of $Ag_2CrO_4 = 10.71$ and $pK_{sp}$ of AgCl = 9.52. (Assuming ionic strength = 0.1)

$$[Ag^+][Cl^-] = 10^{-9.52}$$

$$[Ag^+]^2[CrO_4^{2-}] = 10^{-10.71}$$

$$\frac{[CrO_4^-]}{[Cl^-]^2} - \frac{10^{-10.71}}{(10^{-9.52})^2} - 10^{+8.33} \tag{6-11}$$

(a) Since $[CrO_4^+] = 1.0 \times 10^{-4}$ and $[Cl^-] = 1.0 \times 10^{-7}$, the ion ratio $[CrO_4^-]/[Cl^-]^2 = 10^{+10}$. Since this value is greater than $10^{+8.33}$, $Ag_2CrO_4$ will precipitate first.

(b) Here, $[CrO_4^-] = 1.0 \times 10^{-2}$ and $[Cl^-] = 1.0 \times 10^{-5}$ and the ion ratio is $10^{+8.0}$ which is less than the value $10^{+8.33}$ required for simultaneous precipitation. Hence in this solution AgCl will precipitate first.

## CONDITIONAL SOLUBILITY PRODUCT CONSTANTS

We can develop a totally general expression by using an exact analogy to the approach taken for metal complexes (Chapter 5) to deal with side reactions involving either or both the cation and/or the anion. In the solubility product equilibrium expression for $M_aX_b$, simply substitute the equivalent $\alpha C$ expressions for the concentrations of each ion. Then

$$K_{sp} = [M^{n+}]^a[X]^b = \alpha_M^a \alpha_x^b C_M^a C_x^b$$

or

$$K_{sp}' = K_{sp}/\alpha_M^a \alpha_x^b = C_M^a C_x^b \tag{6-12}$$

where $K_{sp}'$ is called the conditional solubility product constant. In this way, the effect of pH and presence of ligands on the solubility of slightly soluble salts can be readily taken into account.

### Side Reactions Involving the Precipitating Anion

When the precipitating anion is a Brønsted base, then its concentration is pH dependent and, therefore, so is the solubility of the precipitate. Let us consider a few examples.

### Calculations When pH is Known

Many metal ions, in fact all but the alkali metal ions, form slightly soluble hydroxides. The concentration of $[OH^-]$, and hence the pH, required for onset of precipitation depends on their $K_{sp}$ values. The solubility of a metal hydroxide (sometimes referred to as a hydrous metal oxide), $M(OH)_n$, expressed as the maximum concentration of metal ion in solution is

$$[M^{n+}] = K_{sp}/[OH^-]^n = K_{sp}[H^+]^n/K_w^n \qquad (6-13)$$

The logarithmic expression of Equation 6-13 is graphically represented for a number of metal hydroxides (Figure 6.2).

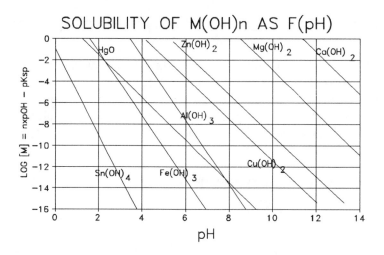

Figure 6.2 Solubility of $M(OH)_n$ as a F(pH)

Notice how the slopes vary with n. Many anions other than $OH^-$ are Brønsted bases. Consider, for example, the solubility of $CaCO_3$ as a function of pH.

**Example 6.6**

When solid $CaCO_3$ is equilibrated with a buffered solution, how does its solubility vary with pH?

$$K_{sp} = [Ca^{2+}][CO_3^{2-}] = C_{Ca}C_{CO3} \cdot \alpha_2$$

$$C_{Ca} = C_{CO3} = (K_{sp}/\alpha_2)^{1/2} \qquad (6-14)$$

The variation of $C_{Ca}$ (the solubility) with pH can be easily plotted as a function of pH using the spreadsheet. First, label **A1** as pH and **B1** as log $C_A$. For **A2** to **A16**, /EF{/BF; /F} 0 to 14 in steps of 1. In cell **B2**, the formula for $(K_{sp}/\alpha_2)^{1/2}$ in spreadsheet format, which is +0.5*@LOG (4.7E-9/(A2^2 + 10^-6.13*A2 + 10^-(6.13+10.33))). Then /EC{/BC; /C} from **B2** to **B2..B16**. For the graph, use **B2** to **B16** as series 1, with **A2** to **A16** as the X series. A figure drawn from this spreadsheet shows that at pH values above $pK_2$ of carbonic acid, the solubility is constant; between $pK_1$ and $pK_2$ where $\alpha_2$ decreases linearly with $[H^+]$ (slope of $\log\alpha_2$ vs. pH is +1) log $C_{Ca}$ increases by 0.5 for every unit decrease in pH; at still lower pH values, log $C_{Ca}$ increases by 1 for every unit pH decreases. If the solution is acid rain at pH 4.0, the spreadsheet will help you find the solubility of a limestone statue (assume pure $CaCO_3$).

## Example 6.7

How does the solubility of $Ca_3(PO_4)_2$ vary in solutions as a function of pH? (Assume that the ionic strength is 0.025 with $pK_{sp}$ = 26.66 and for $H_3PO_4$, $pK_1$ = 2.01, $pK_2$ = 6.93, and $pK_3$ = 11.99.)

$$K_{sp} = [Ca^{2+}]^3[PO_4^{3-}]^2$$

If S = solubility of $Ca_3(PO_4)_2$, then

$$[Ca^{2+}] = 3S \text{ and } [PO_4^{3-}] = 2\alpha_3 S$$

and

$$K'_{sp} = K_{sp}/\alpha_3^2 = 108 \ S^5.$$

Finally,

$$S = (K'_{sp}/108)^{1/5}.$$

A spreadsheet with pH values in 0.2 increments from 2 to 14 placed in column **A**, with the formula for $\alpha_3 S$ in column **B** and that for S in column **C** was used to obtain Figure 6.3. Values of $K'_{sp}$ could also be obtained in a reasonably reliable way by plotting $\log\alpha_3$ vs. pH in the manner described for Figure 4.1.

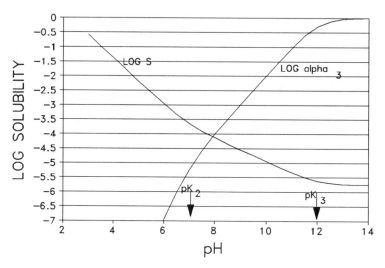

**Figure 6.3 Solubility of $Ca_3PO_4$ as a Function of pH**

## Calculations When pH Not Known

When the pH of the solution is not given or when the pH of the solution is altered by the dissolution of the precipitate, the calculations cannot be carried out without the prior determination of the equilibrium pH.

## Example 6.8

A question of interest to environmental chemists is the solubility of statuary marble in acid rain. This was considered in Example 6.7. How soluble is $CaCO_3$ in "pure" rain water is a related question in which the pH is determined by the $CO_3^{2-}$ of the dissolved $CaCO_3$.

From Equation 6-14,

$$C_{Ca} = C_{CO3} = (K_{sp}/\alpha_2)^{1/2}$$

The PBE of a saturated $CaCO_3$ solution is

$$2[H_2CO_3] + [HCO_3^-] + [H^+] = [OH^-]$$

or

$$(K_{sp}/\alpha_2)^{1/2}(2\alpha_0+\alpha_1) + [H^+] = [OH^-] \qquad (6\text{-}15)$$

Where $(K_{sp}/\alpha_2)^{1/2}$ is substituted for $C_{CO3}$.

This represents a fairly formidable equation which can be "cut to size", however, with the help of the **Pointer** function. The spreadsheet is organized using Columns **A to J** labeled pH, $[H^+]$, $[OH^-]$, $\alpha_0$, $\alpha_1$, $\alpha_2$, $[H_2CO_3](=(K_{sp}/\alpha_2)^{1/2}\alpha_0)$, $[HCO_3^-]$ $(=(K_{sp}/\alpha_2)^{1/2}\alpha_1)$, $F([H])(= (K_{sp}/\alpha_2)^{1/2}(-2\alpha_0+\alpha_1) + [H^+] - [OH^-])$, and Pointer(@log(@abs(F([H])))). Since we know the pH will be above 7, but below 12, /EF{/BF; /F} for pH from 7 to 12 in steps of 0.2. Apply the appropriate formulas for the rest of the columns using $pKa_1 = 6.13$ and $pKa_2 = 10.33$ for carbonic acid.

Plot an xy graph with the pointer as series 1 vs. the pH as the x series. From this graph, the pH can be seen to be about 10.0 - 10.2. Add additional pH values between these limits using 0.01 increments. Finally, using the Sort function, to rearrange the new values according to pH, the pH of the saturated $CaCO_3$ is seen to be 10.13 (Figure 6.4), permitting us to estimate the solubility of $CaCO_3$ to be $(10^{-7.73}/\alpha_2)^{0.5}$, which is $10^{-4.06}$ M, since $\alpha_2$ at this pH is 0.389. The bicarbonate ion is the predominant (61.1%) but not the only significant carbonate species.

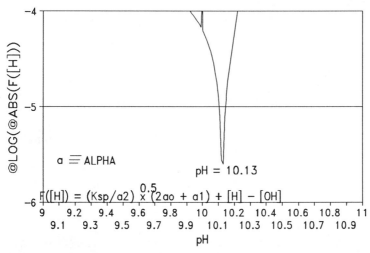

pH OF SATD. CaCO3 IN WATER
USE OF POINTER FUNCTION

**Figure 6.4 pH of saturated CaCO₃**

Let us now consider cases where the pH of the saturated solution indicates that only a very small concentration of the precipitating anion remains. As will be seen, the change of the pH from that of water will depend on the $K_{sp}$. MnS, with a relatively high $K_{sp}$, and CuS, with one that is much lower, will be used to illustrate this effect.

## Example 6.9

Calculate the solubility of MnS in water. Assume that $K_{sp}$ for MnS = $1.0 \times 10^{-11}$; $K_1 = 1.0 \times 10^{-7}$ and $K_2 = 1.3 \times 10^{-13}$ for $H_2S$ and $K_w = 1.0 \times 10^{-14}$.

$$K_{sp} - [Mn^{++}][S^-] - S \cdot \alpha_2 S$$

where S is the solubility of MnS and $\alpha_2$ is the fraction of sulfide ion. Once again, the PBE in this solution is

$$(K_{sp}/\alpha_2)(2\alpha_o + \alpha_1) + [H^+] = [OH^-]$$

from which the spreadsheet graph of the pointer function plotted against the pH may be derived as in Example 6.8 to give the pH of saturated MnS as 9.95. The $\alpha_2$ at this pH is $10^{-2.92}$ and the solubility of MnS is $10^{-4.04}$ M. Further, at this pH, it is readily seen that the predominant (really, the only significant) sulfide species is $HS^-$. This means that for every mole of MnS that dissolves, one mole of $OH^-$ is formed according to the equation,

$$MnS + H_2O \rightleftharpoons Mn^{++} + SH^- + OH^-$$

What happens when the acid used to adjust pH includes the precipitating anion?

## Example 6.10

What is the solubility of barium sulfate in sulfuric acid? We will examine the dependence of the solubility of $BaSO_4$ upon the concentration of $H_2SO_4$, $C_a$.

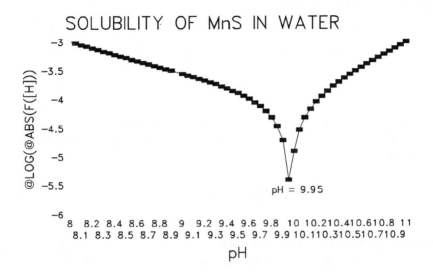

**Figure 6.5 Solubility of MnS in Water**

$$K_{sp} = [Ba^{2+}] [SO_4^{2-}]$$

Proton Balance:  $[H^+] = 2[SO_4^{2-}] + [HSO_4^-]$

Unless concentrated $H_2SO_4$ is involved, $[H_2SO_4]$ is negligible. Substituting for $[HSO_4^-]$ and $[SO_4^{2-}]$ in the PBE,

$$[H^+] = \frac{2 \cdot K_2 \cdot C_a}{[H^+] + K_2} + \frac{[H^+] \cdot C_a}{[H^+] + K_2}$$

(6-16)

$$\text{i.e., } \quad [H^+] = C_a \frac{[H^+] + 2K_2}{[H^+] + K_2}$$

and solving explicitly for [H$^+$],

$$[H^+] = \frac{(C_a - K_2) + \sqrt{C_a^2 + 6C_aK_2 + K_2^2}}{2} \qquad (6\text{-}17)$$

At concentrations where $C_a >> K_2$, [H$^+$] approaches $C_a$. When this happens, the solubility of BaSO$_4$ reaches a constant value, as seen from Equation 6-14,

$$S = K_{sp}/K_2$$

This conclusion presupposes that the ionic strength is essentially constant.

Finally, it should be remarked that often the precipitating agent is referred to as the compound containing the agent, which is the conjugate acid of the precipitating anion. For example, H$_2$S is used in precipitating sulfides. Such precipitation reactions will result in the release of hydrogen ions.

$$Zn^{2+} + H_2S \rightarrow ZnS + 2H^+$$

In the design of suitable buffers for such precipitations, adequate buffer capacity to handle the hydrogen ion released must be employed.

### Side Reactions Involving the Cation

As we saw in Chapter 5, metal ions are Lewis acids and form complexes with a variety of ligands, which can be described as side reactions in precipitate solubility problems. In such problems, the Lewis base or ligand can either be the precipitating anion or another complexing agent.

Let us consider first those situations in which the precipitating anion is also a ligand. Metal ions can form a whole series of coordination or ion-association complexes with such anions The least soluble of the series is the uncharged species, the precipitate.

For example, Pb$^{2+}$ and Cl$^-$ will form PbCl$^+$, PbCl$_2$ (which is slightly soluble in water), PbCl$_3^-$ and PbCl$_4^{2-}$. The solubility of lead in chloride solutions will go through a minimum when the opposing effects of the common ion and formation of anionic complexes counter balance each other. The overall solubility of PbCl$_2$ in a series of solutions conta:ning chloride ions may be quantitatively described in terms of the equilibrium expressions corresponding to the four stepwise complex formation constants and the *intrinsic solubility*, $S_o$, of $PbCl_2$.

Since the solubility of $PbCl_2$, S, is given by the sum of the concentrations of all species containing $Pb^{2+}$,

$$\frac{K_{sp}}{\alpha_{Pb}} - S[Cl^-]^2 \quad or \quad S - \frac{K_{sp}}{\alpha_{Pb}[Cl^-]^2}$$

$$S - [Pb^{2+}] + [PbCl^+] + [PbCl_2] + [PbCl_3^-] + [PbCl_4^{2-}]$$

$$\therefore \quad S - K_{sp} \left[ \frac{1}{[Cl^-]^2} + \frac{\beta_1}{[Cl^-]} + \beta_2 + \beta_3[Cl^-] + \beta_4[Cl^-]^2 \right]$$

$$(6\text{-}18)$$

Note that the third term in the bracket in Equation 6-18 refers to the intrinsic solubility, $S^o$, which has been replaced by its equivalent, $\beta_2 \cdot K_{sp}$. This demonstrates that the solubility product depends not only on the complex formation constants but upon the intrinsic solubility as well.

### Example 6.11

Calculate the solubility of $PbCl_2$ in solutions containing various chloride ion concentrations. The log $\beta_i$ values of $PbCl_4^{2-}$ are 1.29, 2.0, 2.3, 1.7, and log $K_{sp}$ for $PbCl_2$ is -4.78.

A spreadsheet is constructed with values of log [Cl$^-$] from 0 to -5.0 in steps of 0.1 or 0.2 in column **A** and, in successive columns the quantities [Cl], $\alpha_{Pb}$, S, the remaining $\alpha_{PbCl_i}$ values, and the log functions of all of these. From a ligand balance equation:

$$2[PbCl_4^{2-}] + [PbCl_3^-] + [Cl^-] = [PbCl^+] + 2[Pb^{2+}]$$

in which the various $\alpha C$ values are understood to correspond to $\alpha S$ values, a pointer function can be written.

From the resulting spreadsheet, Figure 6.6 was obtained which gives us S as a function of total [Cl$^-$], and a pointer enabling us to obtain the water solubility of $PbCl_2$. According to the pointer function, a saturated solution has a [Cl$^-$] of 0.045 M at which S = 0.025 M.

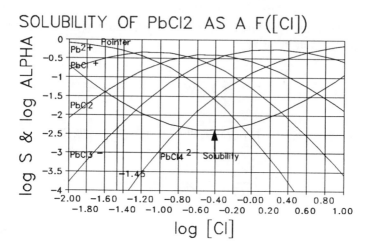

**Figure 6.6 Solubility of PbCl$_2$**

Note that there is a minimum in S at 0.4 M Cl$^-$, reflecting the balance in the common ion effect of chloride ion, which reduces solubility and its role as ligand which, by reducing [Pb$^{2+}$], increases solubility.

## Lewis Bases Other Than the Precipitant

The presence of Lewis bases which form complexes with the metal ion will naturally affect the solubility of slightly soluble metal salts.  For example, ammonia will increase the solubility of silver halides and transition metal sulfides and hydroxides.  Other complexing agents which are of interest in this connection include cyanide, tartrate, citrate, and EDTA ions.

## Example 6.12

Calculate the solubility of AgCl in 0.01 M NH$_3$.  K$_{sp}$ for AgCl = 2.0 x 10$^{-10}$.  The Log $\beta$ values for Ag(NH$_3$)$_2$$^+$ are 3.2 and 7.0.

$$\alpha_{Ag} = [1 + 10^{3.2} \cdot 10^{-2} + 10^{7.0} \cdot 10^{-4}]^{-1}$$

$$\alpha_{Ag} = 10^{-3.0}$$

Therefore,

$$S = \sqrt{\frac{2.0 x 10^{-10}}{10^{-3.0}}} = 4.5 x 10^{-4} \ M \qquad (6\text{-}19)$$

It may be noted that the implicit assumption that the initial concentration of ammonia would not be significantly reduced by either (a) the amount consumed in complexation with $Ag^+$ or (b) the amount transformed to $NH_4^+$, is justified, since neither (a) nor (b) involves as much as 5% of the initial concentration.

As the concentration of $NH_3$ is increased, the solubility of AgCl also increases to the point where it is necessary to correct for the amount of $NH_3$ consumed in the complexation of $Ag^+$. Using the spreadsheet to organize the solution for such a problem should remove any difficulty.

It is not uncommon in precipitation processes of analytical interest that the concentration of the complexing agent as well as that of the precipitating anion is pH dependent. This situation is encountered in the next example.

## Example 6.14

(a) Derive and graphically display an expression for the solubility of CuS in 0.01 M EDTA as a function of pH. Use the following: $pK_{sp}$ of CuS $= 35.1$; Log $\beta_{CuY} = 18.80$; for $H_2S$, $pK_1 = 7.0$ and $pK_2 = 12.9$; for EDTA, $pK_1 = 2.0$, $pK_2 = 2.67$, $pK_3 = 6.16$, and $pK_4 = 10.26$

$$K_{sp} = [Cu^{2+}][S^{2-}] = \alpha_{Cu}C_{Cu}\alpha_2 C_S$$

Then, since $C_{Cu} = C_S$, $S^2 = K_{sp}/\alpha_{Cu}\alpha_2$ where $\alpha_{Cu}$ and $\alpha_2$ have their usual meanings. Hence the solubility, $S$, of CuS in EDTA, as a function of pH is given by

$$S = \sqrt{K_{sp}(1 + \beta_{CuY}\alpha_4 C_Y)/\alpha_2}$$

The spreadsheet used for this calculation starts with filling column **A** with pH values from 0 to 14 in steps of 0.25, then applying suitable formulas to give $[H^+]$, $\alpha_S$, $\alpha_Y$, $\alpha_{Cu}$, $S$, and, finally, log $S$ in the following columns. Although the solubility of CuS in EDTA solution is much greater than it was in water, as might be expected from the high stability of the Cu-EDTA complex, it is still fairly insoluble.

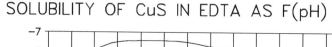

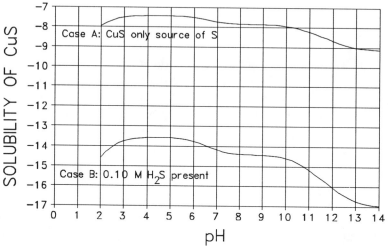

**Figure 6.7 Solubility of CuS in EDTA**

The formation of $Cu(OH)^+$ in this solution has been neglected because the Cu-EDTA complex is much more stable than the $CuOH^+$ complex and furthermore the $[OH^-]$ is much smaller than the $[Y^{4-}]$ over most of the pH range.

Another frequently encountered solubility problem is one in which the excess concentration of the precipitant is kept constant.

**(b)** How does pH affect the $C_{Cu}$ in a solution in which $C_Y = 0.01$ M and $C_{H2S} = 0.10$ M ?      The solubility product expression to that used in (a), except that $[S^{2-}] = \alpha_S(C_{H2S} + S)$, while $[Cu^{2+}]$ is $\alpha_{Cu}S$

$$K_{sp} = \alpha_{Cu}S\alpha_S(C_{H2S}+S)$$

Strictly speaking this represents a quadratic relationship but since S is vanishingly small compared to $C_{H2S}$, the solubility, S, is

$$S \cong \frac{K_{sp}\,\beta_{CuY}\,C_{H_4Y}}{C_{H_2S}\dfrac{\alpha_Y}{\alpha_S}} \qquad (6\text{-}20)$$

## PROBLEMS

**6-1**     Using the Davies Equation, find the value of the ionic strength at which the $K_{sp}$ values differ from the ${}^*K_{sp}$ values by a factor of 10 for the following salts: $AgCl$, $PbCl_2$, $BaSO_4$, $Bi_2S_3$.

**6-2**     What is the solubility of CuS in water? Note that this problem may be solved exactly as the $CaCO_3$ and MnS cases were. Because of the small solubility of CuS, however, whose $K_{sp}$ is $10^{-35.1}$, the pH of a saturated solution is very close to 7.0. In fact, as the pointer graph prepared as described in the previous example shows, the pH is $7.00 \pm 0.01$. Hence, $\alpha_2$ is $10^{-6.1}$, and the solubility of CuS, $(K_{sp}/\alpha_2)^{1/2}$, is $10^{-14.5}$, or $3.2 \cdot 10^{-15}$ M.

**6-3**     In example 6.11, the solubility of $BaSO_4$ was seen to become constant at increased sulfuric acid concentrations. Why doesn't the solubility of silver acetate in acetic acid also become constant with increasing acetic acid concentration?

**6-4**     How does the solubility of $BaSO_4$ in 0.005 M EDTA solution vary with pH?

**6-5**     Derive and display an expression for the solubility of NiS in 0.001 M EDTA as a function of pH. Use the values listed in the Appendices for equilibrium constants.

**6-6**     Repeat calculation for 6.3-6.5 appropriate for solutions that have ionic strengths of 0.1 and use the Davies' equation to obtain equilibrium constants.

# Chapter 7

# Oxidation Reduction Equilibria

---

Oxidation-reduction (**redox**) reactions involve electron transfer and therefore can be used to produce electrical work. This is accomplished by suitably separating the reaction components into two **halfcells** linked in a way that electron transfer must occur through an external circuit. Such a system is called a galvanic cell. This link between electrical and chemical transformations, constituting the field of electrochemistry, is vital to understanding and controlling many phenomena of interest to a wide range of scientists. It provides analytical chemists with powerful methods of monitoring chemical reactions by rapid, precise, and convenient electrochemical measurements of concentrations over a wide range. Transfer of chemical to electrical energy in "batteries" is responsible for an important form of "clean" energy. Electroplating represents another application of wide interest. Electrochemistry has even been exploited as an apocryphal means of accomplishing "cold fusion".

The electromotive force (**emf**) of the cell is a measure, in volts, of the driving force of the chemical reaction involved. Although one cannot measure directly the voltage of each of the two halfcells, the emf of the cell is the voltage difference between the two halfcells. As will be described in detail below, halfcell potentials can be quantitatively evaluated by adopting one halfcell as a standard and measuring all the others by the potential of the cell, $E_x - E_{ref}$. The equilibrium constant of the chemical reaction is another measure of this driving force. The electromotive force of the cell and the equilibrium constant of the chemical reaction involved are therefore related as will be shown below. As a matter of common practice, however, redox equilibria are more often characterized in terms of electromotive force values rather than equilibrium constants.

In redox reactions, to an extent greater than found in other types of reactions, systems which are not reversible or are not in equilibrium are encountered. In such systems, although the predicted potential differences based on equilibrium behavior will not agree with observed values, such calculations are nevertheless useful in describing the relative strengths of oxidants and reductants.

## HALF-CELL REACTIONS

Because electrons are transferred, not created or destroyed, in redox reactions, any such reaction may be considered to be a combination of two halfreactions, each involving an oxidation-reduction couple. For example, in the reaction of $Fe^{+3}$ with $Sn^{+2}$,

$$2Fe^{+3} + Sn^{+2} \rightleftharpoons 2Fe^{+2} + Sn^{+4}$$

the two half reactions are

$$Fe^{+3} + e^- \rightleftharpoons Fe^{+2}$$

$$Sn^{+2} \rightleftharpoons Sn^{+4} + 2e^-$$

and the overall reaction obtained by multiplying the first equation by 2 and adding it to the second, involves the transfer of electrons from the reduced form of the tin halfcell to the oxidized form of the Fe halfcell.

Halfcell reactions bear a striking resemblance to the Brønsted acid-base equilibria. In the latter, a proton shuttles back and forth between the acid and base which are called a conjugate acid-base pair. In a redox halfcell, electrons are the particles transferring between the oxidized and reduced forms of a redox couple.

$$\text{Conjugate acid-base pair:} \quad B + H^+ \rightleftharpoons BH^+$$

$$\text{Redox Couple: Ox} + ne^- \rightleftharpoons \text{Red}$$

In the generalized half-cell equation, n represents the number of electrons necessary to transform a species to the next stable, lower oxidation state. In contrast to proton transfer reactions which occur in steps of one proton at a time, many redox reactions involve the simultaneous transfer of several electrons.

## REDOX STOICHIOMETRY

Half-cell equations are of great value in stoichiometric problems as well as in equilibrium situations. It is not necessary, in relating amounts involved in redox reactions, to use the complete overall equation since all of the necessary information (including that for balancing the overall equation) is contained in the half-cell equations. For example, in the reaction between $Fe^{3+}$ and $Sn^{2+}$ mentioned in the section above, the half-cell equations show that the two electrons lost by the $Sn^{2+}$ are transferred to **two $Fe^{3+}$**, each of which can accept only one each. Thus, balancing redox equations consists in matching the numbers of electrons lost with those gained.

Balancing half-cell equations can be accomplished by one of several techniques, but the **ion-electron** method seems the most convenient and generally useful. It is worth noting that all of the uses of balanced equations, namely, stoichiometric and equilibrium calculations, are well served by the use of halfcell equations. The fundamental factor in redox stoichiometry (description of combining power) is the number of electrons transferred (gained or lost) per species. Thus, from the halfcell of $Cr(VI)/Cr(III)$, we see **6e** per $Cr_2O_7^{2-}$, which translates to an equivalent (combining) weight for $K_2Cr_2O_7$ of one sixth of its GMW; because there are 2 $Cr^{3+}$ in the Eq., the equivalent weight of Cr is one third of its GMW, but one sixth of the GMW of $Cr_2(SO_4)_3$.

## Ion-electron Method of Balancing Half-cell Equations

It is necessary to know the species existing in the solution making up the redox couple of interest; the remainder will be obtained by the balancing method.

There are three steps to follow:

1.   Balance all atoms except for O and H.

2.   Balance O by using $H_2O$ for each O needed, and add two $H^+$ on the other side of the equation for each water. The equation is now **chemically** balanced.

3.   Complete the halfcell by **electrically** balancing using electrons **(don't forget the electron is negatively charged!)** so that the equation will have the same net charge on each side.

For example, suppose we wish to balance the following halfcells: $Mn(VII)/Mn(II)$, $Cr(VI)/Cr(III)$, and $C_2H_5OH/CH_3COOH$, each in an acidic medium. No method of balancing will enable us to know the correct structures of the various ions involved: $MnO_4^-$, $Mn^{2+}$, $Cr_2O_7^{2-}$, and $Cr^{3+}$; we must rely on our knowledge of the chemistry of these systems. Once at this point, however, the ion-electron method takes over. Thus,

$MnO_4^- \rightleftharpoons Mn^{2+}$ (Four O needed on the right, eight $H^+$ on the left.)

$8H^+ + MnO_4^- \rightleftharpoons Mn^{2+} + 4H_2O$ (Charge on the left: $7^+$; right: $2^+$.)

$\therefore$ 5e needed on left

The balanced equation:

$$8H^+ + MnO_4^- + 5e \rightleftharpoons Mn^{2+} + 4H_2O$$

Similarly,

$Cr_2O_7^{2-} \rightleftharpoons 2Cr^{3+}$ (7 O needed on the right, 14 $H^+$ on the left.)

$14H^+ + Cr_2O_7^{2-} \rightleftharpoons 2Cr^{3+} + 7H_2O$ (Charge on the left: $12^+$; right: $6^+$.)

$\therefore$ 6e on left

The balanced equation:

$$14H^+ + Cr_2O_7^{2-} + 6e \rightleftharpoons 2Cr^{3+} + 7H_2O$$

The next, $C_2H_5OH \rightleftharpoons CH_3COOH$ (One O needed on the left, 2 $H^+$ on the right.

$C_2H_5OH + H_2O \rightleftharpoons CH_3COOH + 4H^+$ (Charge on the left: 0; right: $4^+$.)

$\therefore$ 4e needed on right.

The balanced equation:

$$C_2H_5OH + H_2O \rightleftharpoons CH_3COOH + 4H^+ + 4e$$

## EXPRESSIONS OF REDOX EQUILIBRIUM

Using the generic half-cell equation $Ox + ne \rightleftharpoons Red$ as a model, redox equilibrium can be described by the equilibrium expression

$$^*K = a_{Red}/a_{Ox}a_e^n$$

whose logarithmic form,

$$pE = (1/n)\log {}^*K + (1/n)\log a_{Ox}/a_{Red} \tag{7-1}$$

or

$$pE = p^*E° + (1/n)\log a_{Ox}/a_{Red}$$

where, $p^*E$, the negative logarithm of $a_e$, the "electron activity", (and $pE°$, the equivalent of $pK$ or $\log\beta$, the value of $pE$ when the logarithmic term is zero), is an interesting analog of the pH; the equilibrium equation is an analog of the acid-base buffer equation. Another, more traditional, expression of redox equilibrium is called the Nernst equation.

## NERNST EQUATION

If an electromotive cell is operated under reversible conditions, the electromotive force, E, observed is related to the free energy change of the reaction, $\Delta G$, since the electrical work is the product of the charge and the voltage, or

$$\Delta G = -n\mathscr{F}E.$$

where n is the number of Faradays of electricity transferred, $\mathscr{F}$ is the Faraday (96,491 coulombs).

The free energy change of the half cell reaction is also related to the activities of the reactants and products (Chapter 2).

$$\Delta G - \Delta G° + RT \ln \frac{a_{Red}}{a_{Ox}} \tag{7-2}$$

Substituting for $\Delta G°$, its equivalent, $-RT\ln{}^*K$, and substituting $n\mathscr{F}E$ for $\Delta G$, we obtain the **Nernst equation:**

$$E - \frac{RT}{n\mathscr{F}}\ln {}^*K - \frac{RT}{n\mathscr{F}}\ln \frac{a_{Red}}{a_{Ox}} \tag{7-3}$$

In equation 7-3 the term $(RT/n\mathscr{F}) \ln K$ can be replaced by $^*E°$, which is known as the standard electromotive force of the cell. $^*E°$ is seen to be the value of the cell electromotive force that would be obtained if all of the substances in the reaction were at unit activity. On the other hand, if the system were at equilibrium, then the value of E as well as $\Delta G$ would be zero.

Converting Equation 7-3 to logarithms of base 10 and using T = $298°K(25°$ C), we now obtain

$$E = {}^*E° + (0.0592/n) \log a_{Ox}/a_{Red} \tag{7-4}$$

which, where both sides are divided by 0.0592, yields Equation 7-1.

Just as a redox equation can be considered as a combination of two half-cell equations, the cell emf can be considered to be the difference between the potentials of the two halfcells. Values of the absolute potential of any single redox couple cannot be measured, but for all practical purposes, the relative values are sufficient.

## Electrode Potentials

The electrode potential of any single redox couple is defined as the electromotive force of a cell consisting of the standard hydrogen electrode and the electrode in question, written in the following manner:

$$Pt, H_2 \mid H^+ \ (a = 1) \mid\mid Ox_1, Red_1 \mid Pt$$

In this diagrammatic representation of the cell, a single vertical line stands for a phase boundary at which a potential difference is taken into account. The double vertical line represents a liquid junction whose potential difference is considered to be small enough to be ignored usually (but for exact calculations, can be evaluated). This cell diagram implies that the overall chemical equation is written with the hydrogen gas acting as a reducing agent. If the cell diagram is written:

$$Pt + Red_1, Ox_1 \mid\mid H^+ \ (a = 1) \mid H_2 \ Pt$$

then the chemical equation is reversed. This cell emf is not called the electrode potential of the electrode in question. This diagrammatic representation illustrates a general method of describing galvanic cells. In general, the cell reaction corresponding to the diagram is written to correspond with the passage of positive electricity through the cell from left to right.

A positive cell emf signifies that the reaction occurs spontaneously as written (i.e., $\Delta G$ is negative).

Thus, the diagram

$$Pt \mid Red_1, Ox_1 \mid\mid Ox_2, Red_2 \mid Pt$$

corresponds to the reaction

$$n_2Red_1 + n_1Ox_2 \rightleftharpoons n_1Red_2 + n_2Ox_1$$

The emf of this cell, E, is given by

$$E = E_{right} - E_{left} = E_2 - E_1 \qquad (7\text{-}5)$$

The relationship between the electrode potential and the Nernst Equation can be developed by considering the emf of the cell:

$$Pt, H_2 \mid H^+ (a = 1) \mid\mid Ox_i, Red_i \mid Pt$$

The half-cell reactions involved are

$$2H^+ + 2e^- \rightleftharpoons H_2 \qquad\qquad Ox_i + ne^- \rightleftharpoons Red_i$$

These two equations combine to give:

$$nH_2 + 2Ox_i \rightleftharpoons 2nH^+ + 2Red_i$$

in which $2n \mathscr{F}$ of electricity have been transferred. The emf of this cell, $E = E_i - E_H = E_i$ since $E_H = 0$, is given by the Nernst equation 7-3.

$$E_i = {}^*E_i^o - \frac{RT}{2n\mathscr{F}} \ln \frac{a_{H^+}^{2n} a_{Redi}^2}{a_{H2}^n a_{Ox_i}^2} \qquad (7\text{-}6)$$

Since, in this cell, $a_{H_2} = a_{H^+} = 1$, the equation reduces to

$$E_i = {}^*E_i^\circ - \frac{RT}{2n\mathscr{F}} \ln \frac{a_{Red_i}^2}{a_{Ox_i}^2} \tag{7-7}$$

Using logarithms to the base 10, substituting numerical values for R, T, and at 25° C, we have

$$E_i = {}^*E_i^\circ + \frac{0.0592}{n} \log \frac{a_{Ox_i}}{a_{Red_i}} \tag{7-8}$$

In this expression, ${}^*E_i^\circ$ is the standard electrode potential, i.e., the value of $E_i$ when the activities of both the reduced and oxidized forms are unity. Tables of ${}^*E^\circ$ values (see Appendix) are useful for the selection of reductants of suitable strength. It will be noticed that for redox couples having positive ${}^*E^\circ$ values, the oxidized form is superior to the hydrogen ion as an oxidant. The more effective reducing agents are the reduced forms of couples having negative ${}^*E^\circ$ values.

Not all ${}^*E^\circ$ values are obtained from galvanic cell measurements. Difficulties in the form of slow rates of equilibration and the presence of reaction intermediates or extremely reactive components prevent the direct determination of ${}^*E^\circ$ values in such systems as $Cr_2O_7^{2-}$, $Cr^{3+}$ and $Na^+$, Na. ${}^*E^\circ$ values in such systems may be calculated form appropriate thermodynamic data.

A frequent but avoidable error made in using the Nernst equation is the reversal of the ratio of activities. This can be avoided by remembering that increasing the activity of the oxidized form should result in an increase of $E_i$ since this represents an increase in the effectiveness of the couple as an oxidant. Hence, in equations such as Equation 7-5, $a_{Ox_i}$ must be in the numerator.

From this point on, we will be using $E^\circ$, not ${}^*E^\circ$, and concentrations rather than activities. Recasting equation 7-8,

$$E_i = {}^*E_i^\circ + 0.0592/n \, \log[Ox]_i/[Red]_i + 0.0592/n \, \log \gamma_{oxi}/\gamma_{Redi}$$

The values of $E°$ then can be seen to be

$$E_i° = {}^*E_i° + (0.0592/n)\log\gamma_{Oxi}/\gamma_{Redi} \qquad (7\text{-}8a)$$

and, similarly

$$pE = p{}^*E° + (1/n)\log\gamma_{Oxi}/\gamma_{Redi} \qquad (7\text{-}8b)$$

Values of $E_i°$ can be readily calculated when the ionic strengths are describable.

## TYPES OF HALF-CELL REACTIONS

The wide variety of electrode systems or halfcells of analytical importance can be classified in terms of the nature of the components of the redox couple. Of particular importance is the metal that is used as an electron conductor.

If the redox couple involves a metal, then the halfcell consists of a metal immersed in a solution containing the metal ions, e.g.,

$$Zn^{2+} + 2e^- \rightleftharpoons Zn$$
$$Ag^+ + e^- \rightleftharpoons Ag$$

The form of Equation 7-8b that applies in these cases is

$$pE_{Zn} - pE_{Zn}^o + (1/2)\log[Zn^{2+}] \qquad (7\text{-}9)$$

and

$$pE_{Ag} - pE_{Ag}^o + \log[Ag^+] \qquad (7\text{-}10)$$

Metal electrodes corresponding to these halfcells are referred to as
**electrodes of the first kind**. The activity of the metal, Zn or Ag, is unity,
as will be that of every solid component (as well as $H_2O$) of a half-cell
reaction.

A useful variant of this type of half-cell is obtained by coating the metal
with a slightly soluble salt or oxide.

$$AgCl(s) + e^- \rightleftharpoons Ag + Cl^-$$

$$PbO_2(s) + 4H^+ + 4e^- \rightleftharpoons Pb + 2H_2O$$

The Nernst equations for these two electrode systems take the form:

$$pE = pE^\circ + \log(1/[Cl^-]) \tag{7-11}$$

or

$$pE = pE^\circ + pCl \tag{7-12}$$

and

$$pE_{Pb} = pE_{Pb}^\circ + (1/4)\log[H^+]^4 \tag{7-13}$$

or

$$pE_{Pb} = pE^\circ_{Pb} - pH \tag{7-14}$$

The presence of the slightly soluble substance changes the electrode
system from one whose potential responds to changes in the activity of
metal ion to one whose potential depends on the activity of another ion.
These are called **electrodes of the second kind**. It is of analytical importance to be able to follow changes in $a_{H+}$ (for example, in pH measurements) or $a_{Cl^-}$ by means of these or similar electrode systems.

If the halfcell equation couple does not include a metal as one of its
components, then an inert metal such as platinum must be used as the
necessary electron conductor to constitute it as an electrode.  In such
systems the platinum metal does not appear either in the halfcell or in the
Nernst equation.

Consider, for example, the half-cell reactions

$$Ce^{+4} + e^- \rightleftharpoons Ce^{+3}$$

$$2H^+ + 2e^- \rightleftharpoons H_2 \text{ (1 atmosphere)}$$

$$Cr_2O_7^{2-} + 14H^+ + 6e^- \rightleftharpoons 2Cr^{3+} + 7H_2O$$

The corresponding expressions for the electrode potentials are

$$pE_{Ce} = pE°_{Ce} + \log([Ce^{4+}/[Ce]^{3+})$$ (7-15)

For the hydrogen electrode,

$$pE_H = pE°_H + (1/2)\log([H^+]^2/p_{H2})$$

Since $E°_{H+,H2} = 0$ and $p_{H2} = 1$ at 1 atmosphere pressure

$$pE_H = - pH$$ (7-16)

$$pE_{Cr} - pE°_{Cr} + \frac{1}{6}\log\frac{[Cr_2O_7^{2-}][H+]^{14}}{[Cr^{3+}]^2}$$ (7-17)

While the pE° values of the electrodes just discussed are usually found in tables, we should recognize that they can be derived from the combination of the pE°(M$^{n+}$/M°) and the relevant equilibrium constants. For example, the pE° for the AgCl/Ag electrode can be obtained from the pE° of Ag$^+$/Ag and the $K_{sp}$ for AgCl. First,

$$pE = pE° + \log [Ag^+]$$

then,

$$K_{sp} = [Ag^+][Cl^-] \text{ or } [Ag^+] = K_{sp}/[Cl^-].$$

This results in

$$pE = pE° + \log K_{sp}/[Cl^-]$$

Finally, we see

$$pE°(AgCl/Ag) = pE°(Ag^+/Ag) + \log K_{sp}$$

The same consideration applies when acid-base and/or metal complex formation is involved.

### Example 7.1

What is the pE° value of the Fe(III)/Fe(II) couple in a solution containing EDTA at a pH such that FeY$^-$ and FeY$^{2-}$ are the major species? The

$pE°(Fe^{3+}/Fe^{2+}) = 13.04; \log \beta_{Fe(III)Y} = 25.1; \log \beta_{Fe(II)Y} = 14.3.$

We can write the $Fe^{3+}/Fe^{2+}$ equation

$$pE = 13.04 + \log [Fe^{3+}]/[Fe^{2+}]$$

and substitute

$$[Fe^{3+}]/[Fe^{2+}] = \{[FeY^-]/10^{25.1}[Y]\}/\{[FeY^{2-}]/10^{14.3}[Y]\}$$

or

$$pE = 2.2 + \log [FeY^-]/[FeY^{2-}]$$

This example illustrates how dramatically the pE° can change. As the FeY⁻ is much more stable than its $Fe^{2+}$ counterpart, addition of EDTA has increased the reducing power of the system. Adding EDTA to $Fe^{2+}$ solution will cause it to reduce atmospheric $O_2$!

## Electrode Potential Sign Conventions

A great deal of confusion exists concerning electrode potential sign conventions. Since it is possible to write half-cell reactions either as reductions (electrons on the left-hand side of the equation) or as oxidations (electrons on the right) and further to show cell diagrams with the standard hydrogen reference electrode on either the right- or left-hand side, there exists in principle at least four distinctly different sign conventions.

Lively controversies culminated in a decision by the International Union of Pure and Applied Chemistry (IUPAC) in 1953 to write all half-cell reactions as reductions and with a sign that corresponds to the electrostatic charge on the metal. (Thus, metals more active than H acquire a negative charge with respect to the hydrogen electrode and therefore are given negative values for their electrode potentials.) This sign convention for electrode potentials will be used in this book. It is prudent, when reading a different text or a journal paper, to see how the half-cell equation is written so as to understand the sign convention in the Nernst equation.

## Factors That Affect Electrode Potentials

The electrode potential of any redox couple E will vary with the temperature, which affects the value of both E° as well as the coefficient of the logarithmic term, and with the activities of the oxidized and reduced forms of the couple (Equation 7-4). The temperature variation of $E°$, like its counterpart $K$, will depend on the heat of reaction (see Chapter 2).

Values of the electrode potentials, $E$, may vary with (1) changes in the analytical concentrations of the reaction components, (2) the ionic strength of the solution which will affect values of the activity coefficients, (3) the pH of the solution where hydrolysis or hydroxy complex formation is involved, and (4) the presence of complexing agents other than hydroxide ion which affect the concentrations of uncomplexed oxidized or reduced forms when metal ions are involved.

If the solvent is modified as for example by the addition of ethanol to the aqueous solution, the value of $E$ is likely to change significantly. Modifying the nature of the solvent will affect not only $E$ but also the activities of the components of the redox couple by altering activity coefficients and extents of complexation reactions.

## Effect of Concentration on Electrode Potentials

It is obvious from the Nernst Equation 7-5 that the value of the electrode potential of a redox couple will depend on the values of the activities of the reaction components. This means of course that changes in the concentrations of any of the components will likewise affect the value of the electrode potential. The precise way in which the concentration affects the potential is related to the type of half-cell reaction involved, i.e., in the half-cell reactions

$$Ni^{2+} + e^- \rightleftharpoons Ni$$

and

$$Hg^{2+} + 2e^- \rightleftharpoons Hg$$

the electrode potentials vary with the concentrations of the metal cations. In the following half-cell reactions, however,

$$Fe^{+3} + e^- \rightleftharpoons Fe^{+2}$$

$$Sn^{+4} + 2e^- \rightleftharpoons Sn^{+2}$$

the values of the electrode potentials are independent of the concentration of the individual metal ion, but depend on the ratio of the concentrations of the oxidized and reduced forms.

Another type of concentration dependence occurs in the half-cell reactions:

$$Cr_2O_7^{2-} + 14H^+ + 6e^- \rightleftharpoons 2Cr^{+3} + 7H_2O$$

$$I_3^- + 2e^- \rightleftharpoons 3I^-$$

where the coefficients of the oxidized and reduced forms are not the same. For this reason the concentration factors in the Nernst equation will be raised to different powers. Hence in such cases the electrode potential values will vary with the **absolute values** of the concentrations as well as concentration ratios.

## pE-Log Concentration Graphs

It is useful to visualize the way in which the electrode potential of a redox couple varies with the concentrations of the reaction components. A logarithmic diagram analogous to the type used in acid-base systems can be constructed by using the Nernst equation written in the format of Equation 7-1.

$$pE = pE° + (1/n)\log[Ox]_i/[Red]_i \qquad (7\text{-}18)$$

To follow the analogy further, let us define a set of fractions for the redox system. These will be analogous to the $\alpha$ values used previously for acid-base and metal complexation, but to avoid confusion they will be designated as $f$. In the usual meaning of $\alpha$, the sum of the $f$ values of all of the species in the system equals unity. Thus, in the $Fe^{3+}$-$Fe^{2+}$ system, $f_3 + f_2 = 1$(the subscript denotes the oxidation state.

In a system containing species at several oxidation states such as HClO - $Cl_2$ - Cl⁻, $f_1 + f_0 + f_{-1} = 1$. Again, for the uranium system containing species at +6, +5, +4, and +3, $f_6 + f_5 + f_4 + f_3 = 1$.

To obtain $f$ values, let us express Equation 7-18 as its antilogarithm.

$$[Ox]_i/[Red]_i = 10^{n(pE\text{-}pE°)} = (\pi/K)^n \qquad (7\text{-}19)$$

where $\pi = 10^{pE}$ and $K = 10^{pE°}$.  Rearranging Equation 7-19,

$$[0x]_i = [Red]_i(\pi/K)^n. \qquad (7\text{-}20)$$

For the $Fe^{3+}$ - $Fe^{2+}$ system this leads to

$$[Fe^{3+}] = (\pi/K_{32})[Fe^{2+}].$$

Hence,

$$\mathbf{f_3} = [Fe^{3+}]/([Fe^{3+}] + [Fe^{2+}]) = \pi/K_{32}(\pi/K_{32} + 1)$$

$[Fe^{2+}]$ has been factored out since it is common to both numerator and denominator.  Finally,

$$\mathbf{f_3} = \pi/(\pi + K_{32}) \text{ and } \mathbf{f_2} = K_{32}/(\pi + K_{32}) \qquad (7\text{-}21)$$

Compare the forms of $\mathbf{f}$s in Equation 7-21 with the $\alpha$s for the acetic acid system Equation 4-16.

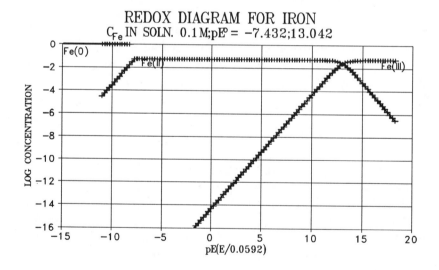

**Figure 7.1 pE-Log C Diagram for Iron**

## Example 7.2

Develop a pE-logC diagram for Fe for a solution of 0.05 M total iron. Consider the $Fe^{3+}$ - $Fe^{2+}$ - $Fe^0$ reactions.

Figure 7.1 was constructed with the help of the spreadsheet. In column A, fill cells with values of pE from -20 to +24 in steps of 1. In succeeding columns label as $[Fe^{3+}]$, $[Fe^{2+}]$ and their logarithmic values. $pE°_{32} = 13.02$ and $pE°_{20} = -7.43$.

For pE values > -7.43,

$$[Fe^{3+}] = 0.05f_3 = 0.05\pi/(\pi + K_{32})$$

and

$$[Fe^{2+}] = 0.05K_{32}/(\pi + K_{32})$$

At pE = -7.43, $[Fe^{2+}] = 0.05$, and, for pE values < -7.43,

$$pE = -7.43 + \frac{1}{2}\log[Fe^{2+}], \therefore [Fe^{2+}] = 0.05(\pi/K_{20})^2$$

We have represented $[Fe^{3+}]$ and $[Fe^{2+}]$ by $0.05f_3$ and $0.05f_2$, respectively,

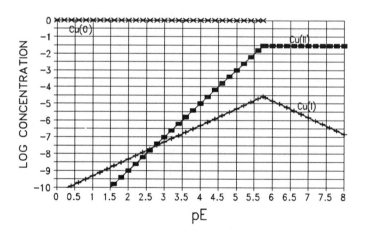

**Figure 7.2 pE - Log C for Copper System**

at pE values above $pE_{20}$. When pE goes below this point, however, C no longer remains at 0.05 M and log $[Fe^{2+}]$, as the predominant aqueous iron species, decreases with a slope of 2 vs. pE, starting at -1.3.

In Figure 7.2 the copper redox system is displayed. This differs in appearance from that of the iron system but the means of developing this diagram are very similar. The equations used are pE = 5.74 + ½log $[Cu^{2+}]$; pE=2.69 +log $[Cu^{2+}]/[Cu^+]$. At pE > 5.74, $[Cu^{2+}]$ = 0.03. At lower values, $[Cu^{2+}]$ = $0.03(\pi/K_{20})^2$. At any pE, $[Cu^+]$ = $K_{21}[Cu^{2+}]/\pi$. The diagram allows us to realize that $Cu^+$ will disproportionate to $Cu^{2+}$ and Cu metal.

In the next example, the system is somewhat more complex, since four oxidation states are represented.

## Example 7.3

Develop a pE-logC diagram for an uranium solution, $C_U$ = 0.01, maintained at pH 0, considering the species from U(VI) to U(III) using the following values: $pE°_{65}$ = 2.70; $pE°_{54}$ = 6.42; $pE°_{43}$ = -8.78.

In this case, U(VI) is present as uranyl ion, $UO_2^{2+}$ which would require $H^+$ to balance the half-cell equation. With pH at 0, however, the calculations are simplified. As before:

$$[U(IV)] = [U(III)](\pi/K43]$$

$$[U(V)] = U(IV)(\pi/K_{54}) = [U(III)](\pi^2/K_{54}K_{43})$$

$$[U(VI)] = [U(V)] (\pi/K_{65}) = [U(III)](\pi^3/K_{65}K_{54}K_{43})$$

These f values of redox systems are closely analogous to those of acid-base systems. When the species in the system represent oxidation states varying by one, the simple rules for writing **f** expressions are:

$$C_U - [U(III)]\{1 + \frac{\pi}{K_{43}} + \frac{\pi^2}{K_{54}K_{43}} + \frac{\pi^3}{K_{65}K_{54}K_{43}}\}$$

OR

$$f_3 = \frac{[U(III)]}{C_U} - \frac{\pi^3}{\pi^3 + K_{65}\pi^2 + K_{65}K_{54}\pi + K_{65}K_{54}K_{43}} - \frac{\pi^3}{D}$$

$$f_4 - K_{65}\frac{\pi^2}{D} \quad f_5 - K_{65}K_{54}\frac{\pi}{D} \quad f_6 - \frac{K_{65}K_{54}K_{43}}{D}$$

(a)    The denominator for each **f** in a redox system of n adjacent oxidation states is identical; write it first.

(b)    The denominator is a decreasing power series in $\pi$, starting with $\pi^n$ and finishing with $\pi^0$ or 1. In each term after the first, an additional constant becomes a factor. The denominator will have **n + 1** terms. The second term will have a single K; the third, a product of two Ks; the last will be a product of **n** K values.

(c)    Since each term in the denominator is proportional to the concentration of one of the species in the redox system, select the appropriate term for the numerator. Thus, the **f** for the most oxidized species has $\pi^n$ in the numerator. The **f** for the species having the lowest oxidation state will have the product of n K values in its numerator.

To draw the pE-log C diagram we proceed in much the same manner as before. Column **A** is used to step off pE from -14 to +12 in steps of 0.1. Corresponding values $\pi$, $f_6$, $f_5$, $f_4$, $f_3$, and log f values for the four are calculated and filled in columns B to **J** in the usual manner. The results, in Figure 7.3, reveal once again the great advantage of the spreadsheet approach in clarifying even complex equilibrium systems.

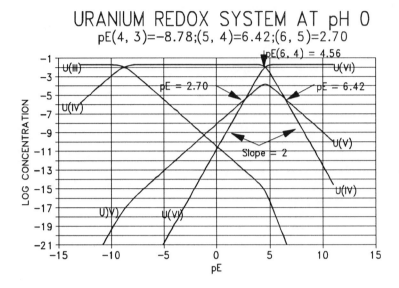

## URANIUM REDOX SYSTEM AT pH 0
### pE(4, 3)=−8.78;(5, 4)=6.42;(6, 5)=2.70

**Figure 7.3 pE - Log C for Uranium System**

The diagram reveals that as the pE increases, the predominant species pass from U(III), to U(IV), and finally U(VI). U(V) is, at best, a minor component; if one starts with U(V) in solution it would disproportionate into U(IV) and U(VI). The value of $pE^\circ$ (6,4), intentionally omitted from the information furnished with Example 7-3, is obtained from the graph. Its value, 4.56, coincides with the tabulated value. While this is required by thermodynamics, it is nevertheless gratifying to obtain without any effort. Note also that the slopes of the log U(IV) and log U(VI) lines are two, as required by the corresponding Nernst equation. A further note of interest: the intersection of the lines for U(IV) and U(V) occur at pE = 6.42 and that for U(V) and U(VI) do so at pE = 2.70 as expected from the $pE^\circ$ values. The highest [U(V)] occurs pE = 4.56, where [U(IV)] = [U(VI)], corresponding to about 1% of $C_U$.

The vanadium system V(III) to V(V) differs from that of uranium in that there is a distinct pE region in which each of the four species predominates (Figure 7-3) values of $pE^\circ$ (5,3) and $pE^\circ$ (5,2), available in the literature, are reliably obtained as the appropriate intersections marked on the diagram. The diagram also marks $pE^\circ$ (4,2) which is not usually shown in the literature.

Diagrams of pE - log C can be constructed for systems which, by traditional methods, would have been far beyond the scope of this text. For

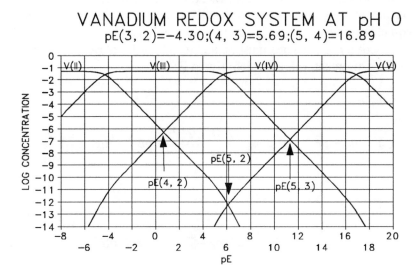

**Figure 7.4 pE - Log C for Vanadium System**

students and professional chemists alike, this application of spreadsheet graphics will make such visual representation of complex redox systems more accessible and certainly more understandable.

## FORMAL POTENTIALS, $E°$ ($pE°'$)

When we considered metal complex formation equilibria, it was found very convenient to introduce the parameter, the conditional formation constant $\beta'$. The use of $\beta'$ simplified the proper handling of side reactions, including proton transfer reactions of the ligand, complexation of the metal with other ligands present in the solution, and even those cases where the primary metal complex was itself involved. With redox equilibria there is also such a parameter, called the **formal potential**, $E°'$, enabling us to write the Nernst equation as

$$E = E°' + (0.0592/n) \log C_{Ox}/C_{Red} \qquad (7\text{-}22)$$

or

$$pE = pE°' + 1/n \log C_{Ox}/C_{Red}$$

If the half-cell equation includes $H^+$, then $pE^{o'}$ will depend on pH. The formal potential, will also incorporate the $\alpha_M$ values relating to the presence of complexing ligands and $\alpha$, values if the ligand is affected by pH. Finally, activity coefficients must be included in order to relate to potentiometric measurement values which give $p^*E^{o'}$.

For example,

$$(a) \quad pE - pE^o + (1/5)\log\frac{[MnO_4^-][H^+]^8}{[Mn^{2+}]} - pE^{o'} + 1/5\log\frac{C_{Mn(11)}}{C_{Mn(11)}}$$

where

$$pE^{o'} - pE^o + (1/5)\log\frac{\alpha_{MnO4^-}}{\alpha_{Mn}} - (8/5)\ pH$$

$$(b) \quad pE - pE^o + \log\frac{[Fe^{3+}]}{[Fe^{2+}]} - pE^{o'} + \log\frac{C_{Fe}(III)}{C_{Fe}(II)}$$

$$where \quad pE^{o'} - pE^o + \log\frac{\alpha_{Fe^{3+}}}{\alpha_{Fe^{2+}}}$$

(c) $\quad pE = pE^o + \frac{1}{2}\log [Cu^{2+}] = pE^{o'} + \frac{1}{2}\log C_{Cu}^{2+}$

where

$$pE^{o'} = pE^o + \frac{1}{2}\log \alpha_{Cu}$$

## POURBAIX DIAGRAMS

A great deal of information about redox systems can be conveyed efficiently by Pourbaix[*] diagrams in which $pE^{o'}$ values for various oxidation states are plotted against pH or another relevant variable such as log[L]. The resultant diagrams consist of a series of appropriately drawn line segments. The areas between the lines represent regions in which one

[*]A French electroanalytical chemist

species predominates and are labeled accordingly. Let us consider the effect of pH on $pE_{Ni}^{\circ\prime}$. The following equations will be considered.

(a) $Ni^{2+} + 2e \rightleftharpoons Ni^{\circ}$ $\qquad\qquad\qquad pE^{\circ} = -4.34$

(b) $Ni(OH)_2(s) \rightleftharpoons Ni^{2+} + 2\ OH^-$ $\qquad pKsp = 15.80$

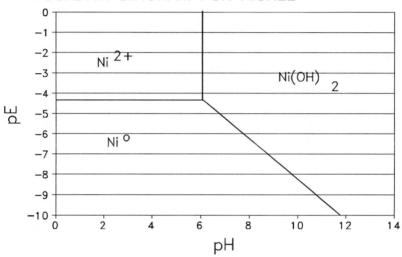

**Figure 7.5 Pourbaix Diagram for Nickel**

The $pE^{\circ\prime}$ will remain unchanged with pH until the value at which $Ni(OH)_2(s)$ begins to precipitate from a solution in which $[Ni^{2+}] = 1$

$$10^{-15.80} = [OH^-]^2 \text{ or } pOH = 15.80/2 = 7.9$$

$$pH = pK_w - pOH = 6.1$$

At $pH = 6.1$, the equation obtained by subtracting (a) from (b) applies:

$$Ni(OH)_2(s) + 2e \rightleftharpoons Ni^{\circ} + 2OH^-$$

The $pE^{\circ\prime}$ change with pOH and therefore, with pH, with a slope of unity because although there is a two-electron change, two $OH^-$ are produced.

Fill column **A** with pH values from 0 to 14 in steps of 1, adding the value of 6.1. If necessary, conduct a sort in **A** to place 6.1 properly. In column **B**, place the value -4.34 from 0 to 6.0. At pH = 6.1 enter the formula, **-4.34 - (pH - 6.1)** using the appropriate **A** cell number for pH. Fill the remainder of **B** cells by the block copy command. Add a vertical line at pH 6.1 for values of $\geq$ -4.34 of $pE^{o\prime}$ (either by hand or, if using QPRO, by the **Annotate** command, **/GA**).

**Example 7.4**

Develop a Pourbaix diagram for $Fe^{3+}/Fe^{2+}$, i.e. $pE^{o\prime}$ vs. pH.

The relevant equations are:

(a)  $Fe^{3+} + e \leftrightarrow Fe^{2+}$               $pE^{o} = 13.04$
(b)  $Fe(OH)_3(s) \leftrightarrow Fe^{3+} + 3OH^-$     $pKsp = 35.22$
(c)  $Fe(OH)_2(s) \leftrightarrow Fe^{2+} + 2OH^-$     $pKsp = 15.10$

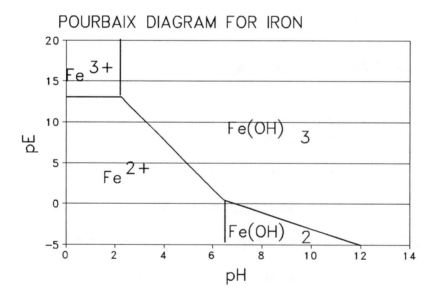

**Figure 7.6 Pourbaix Diagram for Iron**

From (b) and (c) we solve for the pH of the onset of precipitate formation from a 1-M solution. These are 2.26 and 6.45 for (b) and (c), respectively. In filling **A** with pH values from 0 to 14, add the values 2.26 and 6.45. Use the Sort command to place them properly. In column **B**, enter 13.04 and copy it to all cells corresponding to the row in which pH = 2.26. At 2.26, enter the formula +**13.04 - 3(pH - 2.26)**, copying until you reach pH 6.45. From this pH on, enter the formula +**B(x) - (pH - 6.45)** where **x** is the row corresponding to pH = 6.45. Add vertical lines at pH 2.26 and 6.45 to the graph.

The chlorine system presents an interesting case because of the limited pH range of stability of $Cl_2$.

## Example 7.5

Develop Pourbaix diagram for Cl(+1), Cl(0), and Cl(-1) in the pH range 0 to 14.

The relevant equations:

(a) $Cl_2 + 2e \leftrightharpoons 2Cl^-$               pE° = 23.65
(b) $HClO + H^+ + e \leftrightharpoons 1/2\ Cl_2 + H_2O$     pE° =- 27.02
(c) $HClO = H^+ + 2e \leftrightharpoons Cl^- + H_2O$       pE° = 25.34
(d) $HClO \leftrightharpoons H^+ + ClO^-$           pKa = 7.3

First fill **A** with pH values from 0 to 14 in steps of one, including 7.3 property through sort. In **B2**, enter 23.65 and block copy (**/BC**) this value to the cell corresponding to 6. In **C2**, enter the formula +**27.02 - pH** and continue this line to the cell corresponding to pH 6. Develop an XY graph using **B2-B8**, and **C2-C8**, as the series. View the graph and locate precisely the pH where the lines intersect (23.65 = 27.02 - pH), pH = 3.37. Enter this value in **A** properly through Sort. Return to column **B** and delete values corresponding to all pH > 3.37. Likewise in the cell of C at pH = 3.37 enter the formula +**27.02-1/2pH**. Fill all C cells to pH = 7.3. The cell corresponding to pH = 7.3 should have the pE value 19.72. In this cell enter the formula +**25.34 - pH** and copy for the cells of remaining pH range.

## Effect of Complex Formation on Electrode Potentials

Redox diagrams of this sort can also be used to represent the effect of complexation on formal potentials. Figure 7.8 represents the effect of

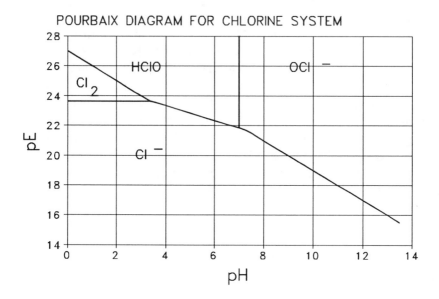

**Figure 7.7 Pourbaix Diagram for Chlorine**

chloride ions on the $pE^{o'}$ of the $Fe^{3+}$-$Fe^{2+}$ couple. As [$Cl^-$] increases, the $pE^{o'}$ decreases with a slope of -1, when $FeCl^{2+}$ predominates. When the $FeCl_2^+$ becomes the major species, the slope changes to -2. The spreadsheet is constructed with log [$Cl^-$] in **A** ranging from -4 to +1 in steps of 0.25 and pE in **B**. The value 13.04 is entered in **B2** and block copied to the cell corresponding to log [$Cl^-$] = -1.48(because log $\beta_{FeCl}$ =1.48). At this point, the formula **13.04 - (log [Cl] -(-1.48))** is used for pE through log [$Cl^-$] = -0.65 (the value at which $FeCl_2^+$ and the monochloride complex are equal). Now the formula **+B(x) - 2\*(log [Cl] -0.65)** , where x represents the cell number corresponding to that in **A** containing -0.65, is entered and copied for the remaining **B** cells.

One of the methods by which the formation constants of metal complexes can be determined involves the measurement of electrode potentials of half cells of the type $M^{n+}$ + $ne^-$ ⇌ M in a series of solution of varying ligand concentrations.

The material presented in this chapter is based on the assumption that achievement of equilibrium is rapid. It must be pointed out, however, that redox processes are sometimes quite slow.

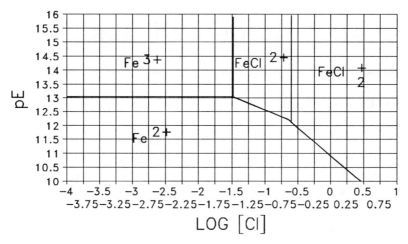

POURBAIX DIAGRAM Fe(III)−Fe(II)
IN CHLORIDE MEDIUM

**Figure 7.8 Pourbaix Diagram for Fe in Chloride Medium**

## PROBLEMS

7-1   a)   Develop the diagram for pE°′ vs. pH for the HClO, Cl$_2$, Cl⁻ System

      b)   Using appropriate pE°′ values from (a), develop pE vs. logC for this system at pH 0, 5, and 8, using different line styles to different distinguished results at each pH (If QPRO 3.0 is available, use the 3D graphical option for 2 or 3 pH values).

7-2   Develop the redox diagram (pE vs. log [Cl⁻]) for the copper system in a chloride medium. Place vertical lines in this diagram (for processes that are independent of E) in an analogous manner to the vertical lines in Figure 7-3 and 7-4. These represent the log $\beta_{CuCl_i}$ values. The vertical line corresponding to solid CuCl would be placed at pCl = 6.7, if it were not for the fact that Cu(I) disproportionates to Cu$^{+2}$ and Cu° at the lower pCl value of 3.9. The horizontal lines involve redox processes in which both oxidized and reduced forms contain the same number of complexed chlorides.

7-3   Is it easier to oxidize Fe° in a solution of pH5 if (a) NOH$_2$S is present or (b) if [H$_2$S] = 5 X 10⁻⁴M?

# Chapter 8

# Titrations I
# Brønsted Acids and Bases

## Nature of Titrations

Titrimetry is an analytical process involving adding portions of a reagent in solution, called a **titrant**, to react, in a specific and well-defined manner, with a component of an analyte solution, until an exact equivalent (i.e., when the titration reaction is exactly complete) of the reagent has been added. This is called the **equivalence point**. In this respect, titrimetric analysis differs from most other types of analysis in which an excess of reagent is added. Polyfunctional species will lead to more than one equivalence point in a titration.

In this respect, titrimetric analysis differs from most other types of analysis in which we add an excess of reagent. Titrimetric methods can be classified according to the type of chemical reaction involved, such as acid-base, precipitation, metal complex formation (complexometric), oxidation-reduction, among others.

In order for a useful, reliable titration to be achieved, the titration reaction must go to completion in a relatively rapid fashion and be free of any interfering side reactions. The substance used as titrant must be readily available in pure, stable form and be reasonable in cost. Further, it must be possible to readily detect the **endpoint**, the experimental approximation of the equivalence point, by suitable **visual** or **instrumental** indicators.

A visual indicator is generally a highly conjugated aromatic organic compound having the characteristic reactivity of the titration type (i.e., it may be a weak acid, metal complexing agent, reductant, etc.) whose coupled forms (HIn/In, MIn/In, Ox/Red) are highly and differently colored. The visual indication of the endpoint depends on observing in a narrow concentration range of the critical variable, the color change of the indicator from one form to the other. Use of an instrumental indicator, e.g., a pH meter, not only gives a reasonably quantitative measure of the critical variable concentration at the endpoint, but obviously, at every point along the titration curve as well.

When the aim of the titration is simply to perform an analytical determination, either visual or instrumental indicators will serve the purpose. **Titration errors**, which are the differences between experimentally determined endpoints and ideal equivalence points, are generally larger and less reproducible for visual endpoints than for instrumental endpoints. Further, only with an instrumental indicator can a **titration curve** be described. Therefore, instrumental indicators are more suited to the study of the fundamentals of titration processes.

A **titration curve** is a graphical representation of the relation of the critical variable, e.g., $[H^+]$ or pH, $[M^{n+}]$ or pM, usually plotted on the Y axis, to the volume of titrant added or some function derived from this volume, such as F, the fraction titrated (moles titrant per mole substance titrated), plotted on the X axis. Titration curves are called **LINEAR** or **LOGARITHMIC**, depending on the way the concentration of the critical variable is expressed. Current sensor instrumentation is sufficiently reliable so that linear and logarithmic plots can be readily interchanged.

A linear plot has the advantage of using titration points not too close to the endpoint, so that the painstaking dropwise search for the endpoint is unnecessary, unlike the common practice with logarithmic titrations. Further, precise algebraic calculations of the endpoint can be carried out readily. Both of these types are useful. The linear curve is particularly practical when dilute solutions are titrated. Logarithmic curves are convenient when the concentration of the critical variable changes by factors of more than a thousand. The manner of locating endpoints also differs in these two types of titration curves as will be explained below.

Traditionally, titration curve calculations are described in terms of equations that are valid only for parts of the titration. Equations will be developed here that reliably describe the entire curve. This will be done first for acid-base titration curves. In following chapters, titration curves for other reaction systems (metal complexation, redox, precipitation) will be developed and characterized in a similar fashion. For all, graphical and algebraic means of locating the endpoints will be described, colorimetric indicators and how they function will be explained, and the application of these considerations to (1) calculation of titration errors, (2) buffer design and evaluation, (3) sharpness of titrations, and finally, (4) in Chapter 18, the use of titration curve data to the determination of equilibrium constants will be presented.

In acid-base titration, the appropriate concentration variable is $[H^+]$ or, most commonly, pH. We have already developed all the necessary equations earlier (Chapter 4). The only difference is that now, instead of writing a proton balance equation (PBE) for a single set of conditions, these apply to a whole family of points, i.e., those involved in the entire titration.

## Use of PBE to Derive Titration Curves

**1. Strong acid - strong base.** Let us add a specified volume, $V_A$ mL, of a solution of $C_A$ M HCl to a vessel and titrate this with $C_B$ M NaOH. The PBE can be written that describes the titration mixture for any volume, $V_B$ mL, of base that is added:

$$[Na^+] + [H^+] = [OH^-] + [Cl^-]$$

where both $[Na^+]$ and $[Cl^-]$ can be described in terms of the diluted base and acid, i.e.,

$$[Na^+] = V_B C_B/(V_A + V_B) \quad \text{and} \quad [Cl^-] = V_A C_A/(V_A + V_B)$$

to give

$$\frac{V_B C_B}{V_A + V_B} + [H^+] \quad - \quad [OH^-] + \frac{V_A C_A}{V_A + V_B} \tag{8-1}$$

The quantity **F, titration fraction**, defined as the number of moles of titrant added, $C_B V_B$, per mole of analyte acid, $C_A V_A$, is another useful way to describe a titration.

## Example 8.1

Generating a Strong Acid-Base Curve. The titration curve can be generated with the help of Quattro. Start by /EF{/BF; /DF} A2..A30 values of pH from 0 to 14 in steps of 0.5. Then in columns B and C, calculate $[H^+]$ and $[OH^-]$ from pH values in A; in columns D and E, calculate $V_B$ and F. A plot of pH vs. either $V_B$ or F describes the titration curve as seen in Figure 8.1. (Note: the spreadsheet design permits several curves in which the variable in the Y-axis changes with a given X-axis variable. Therefore, Figures 8.1 and 8.2 has pH as its X variable.)

$V_B$ is plotted from 0 to 100 mL on the X axis against pH on the Y axis for 50 mL of various concentrations of HCl with equal concentrations of NaOH as titrant. Notice how the pH changes slowly both well before and well after the region of the equivalence point, and changes very rapidly near the equivalence point. Further, it can be seen that the region of rapid pH

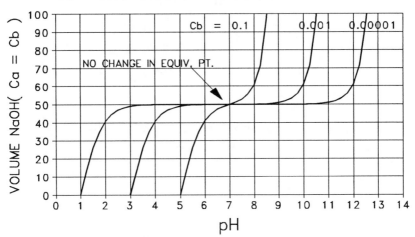

**Figure 8.1 Titration of HCl at Various Concentrations**

change near the endpoint decreases with the acid concentration. The pH of the equivalence point remains essentially the same, since the composition of the solutions at the equivalence points is NaCl of various concentrations.

### 2. Weak acid - strong base

The corresponding PBE when $H_N B$, a weak polyprotic acid, replaces HCl is:

$$[Na^+] + [H^+] - [OH^-] + [H_{(N-1)}B] + 2[H_{(N-2)}B] + N[B]$$

$$or \quad \frac{V_B C_B}{V_A + V_B} + [H^+] - [OH^-] + \frac{V_A C_A}{V_A + V_B} \cdot \sum i \cdot \alpha_i \tag{8-2}$$

$$\text{and} \quad V_B - V_A \cdot \frac{(C_A \sum i\alpha_i - [H] + [OH])}{(C_B + [H] - [OH])}$$

(8-3)

$$\text{or} \quad F - \frac{V_B C_B}{V_A C_A} - \sum i\alpha_i + \frac{V_A + V_B}{V_A C_A} ([OH^-] - [H^+])$$

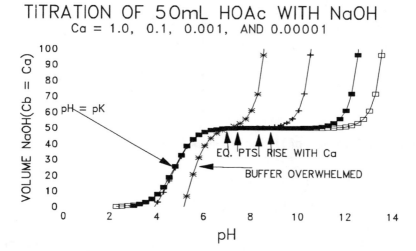

**Figure 8.2 Titration of Acetic Acid at Various Concentrations**

## Example 8.2

Generating the titration curves for weak acids with NaOH. Figure 8.2 (note: here, too, pH is in the unusual position as X-variable) illustrates the titration curves obtained by using Equation 8-2 for titrations of acetic acid of various concentrations with equal concentrations of NaOH.

Notice the S-shaped region (from about 10 to 90% neutralization of the acid) before the equivalence point corresponding to the buffer region (mixture of HOAc and NaOAc). The close resemblance of these segments of the curve demonstrates the independence of buffer pH from absolute acid concentration; when HOAc is very dilute ($10^{-5}$ M), however, its dissociation is essentially complete so that its titration curve is like that of HCl at the same concentration.

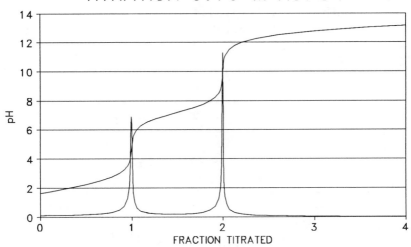

**Figure 8.3 Titration Curve for 0.1 M H₃PO₄ with 0.1 M NaOH**

To further illustrate the utility of Equation 8-3, a titration curve for 0.1 M $H_3PO_4$ has been derived (Figure 8.3). Along with the titration curve itself, the "first derivative" curve has been developed, based on the numerical difference of the change in pH (**A3 -A2**) divided by the corresponding change in F. This gives a measure of the "Sharpness Index" of the titration (see section below). Can you explain why there are only two maxima visible, even though $H_3PO_4$ is a triprotic acid.

The student should develop spreadsheets and similar titration curves for other polyprotic acids such as succinic and citric acids to have as supplement to this text. From such curves  qualitative observations on the ease of defining the individual endpoints can be made.

**3. Weak base - strong acid.** The titration curve equation for a weak base vs. a strong acid can be derived in a manner similar to the cases described above. For $V_B$ mL of $C_B$ M $Na_2CO_3$ titrated with $C_A$ M HCl (Figure 8.4).

Notice that $2\alpha_0 + \alpha_1 = 2 - (\alpha_1 + 2\alpha_2)$ because $2 = 2(\alpha_0 + \alpha_1 + \alpha_2)$ or, more generally for a polyacid base,

$$N - \sum i\alpha_i - \sum (N - i)\alpha_i = \bar{n}_H$$

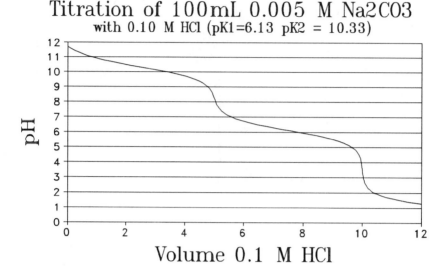

**Figure 8.4  Titration of Na₂CO₃ with HCl**

$$\text{Then } F = \sum_{i}^{N} (N-i)\,\alpha_i + \frac{V_A + V_B}{V_A C_A}\,([H^+] - [OH^-]) \qquad (8\text{-}4)$$

It is noteworthy that the corresponding $f(\alpha)$ parts of the titration curves of the acid, $H_N B, (f(\alpha) = \sum i\alpha_i)$ with NaOH and the base, the totally deprotonated $B(f(\alpha) = \sum (N-i)\alpha_i$ or $\bar{n}_H)$ with HCl are, complementary. The first is a measure of the extent of removal of protons from the acid, and the second, addition of protons to the base. The term, $\bar{n}_H$, represents the average number of **bound** protons per base. The addition of protons to a base is strongly analogous to metal complex formation where, $\bar{n}$, the average number of ligands bound to the metal ion, will be introduced.

## Calculations of pH for Selected Individual Points on the Titration Curve

The equation curve equations developed earlier can be used for single points as well as for entire curves. If only a few separate points are needed, however, calculations designed for each point might be simpler than the titration curve equation. Such calculations involve (a) stoichiometry of the

A/B reaction, (b) recognition of the type of pH situation results from this, and finally, (c) using the appropriate, simple formula (see Problem 4-1) and making the necessary substitutions.

## Example 8.3

What are the pH values of a solution of 50 mL of 0.1 M HX to which $V_B$ = 0, 25, 50, and 75 mL of 0.1 M NaOH have been added?

(a)     $V_B$ = 0. No stoichiometric calculation necessary before any base is added. When $X^-$ is $Cl^-$, the formula for strong acid is used. pH = -log 0.1 = 1.0. When $X^-$ is $OAc^-$, the weak acid formula applies. pH = $(1/2)(pK_a - logC_A)$ = $(1/2)[4.74 - (-1)]$ = 2.87.

(b)     $V_B$ = 25 mL. At this point, 25 x 0.1, or 2.5 mmol of base forms an equal amount of NaX, and (50 x 0.1 - 2.5) or 2.5 mmol HX remain. When $X^-$ is $Cl^-$, $[H+]$ = 2.5/(50 + 25) = 0.03 and pH = 1.52. With $X^-$ as $OAc^-$, the pH is defined by the buffer equation, pH = pK + log[B]/[HB]. Notice that the pH is independent of the solution volume. Here, pH = $pK_a$ = 4.74.

(c)     $V_B$ = 50 mL, the equivalence point. If $X^-$ is $Cl^-$, the pH = pOH = $(1/2)pK_w$ = 7.0. For $X^-$ as $OAc^-$, use the weak base formula, pH = $(1/2)(pK_w + pK_a + log(5.0/50 + 50))$. pH = $(1/2)(14 + 4.74 - 1.30)$ = 8.72.

(d)     $V_B$ = 75 mL. All of the HB has reacted and there is an unreacted 2.5 mmol $OH^-$ in 125 mL of solution, corresponding to 0.02 M. Since $OAc^-$ is a weak base, it will not affect the pH anymore than $Cl^-$ will. pOH = -log(0.02) = 1.70; pH = $pK_w$ - pOH or 12.30 in both cases.

**If this were a diprotic acid titration,** the same methods as used in (a) to (d) would apply. There would of course be two buffer ranges and two equivalence points. Take care that you use the proper $pK_a$ value in the buffer equation. The calculation at the intermediate equivalence point calls for the formula for an amphiprotic salt, pH = $\frac{1}{2}(pK_{a1} + pK_{a2})$.

## CHARACTERIZING TITRATION CURVES

1.     **Sharpness of titrations**. The endpoint of a titration is referred to as **sharp** when the pH change in the vicinity of the endpoint is steep. Therefore, the **sharpness** of the endpoint can be measured as the slope (dpH/dF) of the titration curve near the endpoint. Although the slope is described by an equation obtained by differential calculus, it can be very simply evaluated to a good and useful approximation using $\Delta pH/\Delta F$ with the help of the spreadsheet. For this purpose, it is best to use pH increments no larger than 0.1, particularly in the region of the endpoint. Hence, in column A, use the block, A2..A141, to fill with 0.1 changes in pH. Then in cell, E2, use formula $+(A3 - A2)/(D3 - D2)$. On the graph, use the E2...E141 values as Series 2 {or B, for 123} (Series 1 {or A} being the D2.D141) with A2.A141 as X axis values (Figures 8.3, 8.4). You will note that this sharpness indicator, known as the **sharpness index**, has an appreciable value only in the vicinity of the equivalence point (i.e., with a maximum at the value where **F**, the titration fraction, equals one). In titration curves describing the neutralization of polyprotic acids, there will be a peak in the $dpH/dV_B$ value at each of the equivalence points. The height of each peak is a measure of the sharpness of the equivalence point.

2.     **Buffer Region(s) of the Titration**. If instead of calculating the change in pH per unit change in titration fraction (or volume) described above, the reciprocal, the change in titration fraction per change in pH, $+(D3-D2)/(A3-A2)$, were plotted, we could locate the *flattest* rather than the steepest part of the titration curve. The degree of flatness measures the buffer efficiency or, as it is commonly called, the buffer index.

As seen in Figure 8.4, this function has maxima at F = 0.5(V = 2.5 mL) and F = 1.5 (V = 7.5), which correspond to a 1:1 ratio of untitrated $CO_3^{2-}$ to $HCO_3^-$ to $H_2CO_3$. These equimolar mixtures are the most efficient buffer mixtures of the carbonate system; buffer efficiency decreases both at pH values higher and lower than pH = pK, and helps us understand why the desired pH value of a buffer mixture should be less than one unit away from the $pK_a(s)$ of the system chosen.

Treatment of buffer efficiency in general is not limited to examination of various monoprotic and polyprotic acids, but rather to the behavior of mixtures of acids. Further, from plots of buffer index vs. pH, it is easy

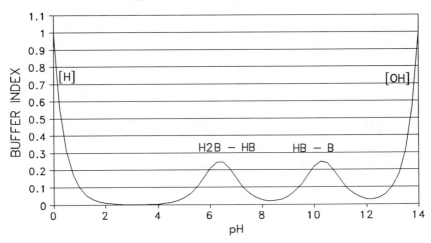

Figure 8.5 Buffer Index in the Carbonate System

to see that strong acids and bases are reasonable buffers for the extreme (low and high) Ph ranges. This leads to the definition of a buffer as a solution that has **neutralization capacity**, rather the more limited but commonly used definition as a mixture of a weak acid and its conjugate base (See Figure 8.5).

3.      **Visual Acid-Base Indicators**. Historically, the earliest titrations were performed using dyes whose color changed as the solution went from acidic to basic condition. An acid-base indicator is a **dye** that is a **weak acid/base pair** with **characteristic spectra** are **distinct from each other**. Because the indicator is itself an A/B system, we must use a very small amount to minimize its interference with titration stoichiometry. Hence, indicators should be highly colored, i.e., **very high** molar absorbances ($\epsilon > 10^4$). For use in aqueous titrations, the indicator molecule must be water soluble at the concentrations employed. Many contain a sulfonic acid group to promote aqueous  solubility.

What is meant by the pH range of an indicator?  How close does $pK_{In}$ need to be to $pH_{eq}$? The concentration ratio of acid to conjugate base forms, $[HIn]/[In](= [H^+]/K_{In})$, determines the color we observe. Consider a two-colored indicator such as methyl orange, in which HIn and In have red and yellow colors. When $[HIn]/[In] = 1$, the color will be orange, i.e., a mix of red and yellow. At this point, the center of the color change range, pH $= pK_{In}$. Colors of mixtures of 1:10 or 10:1 will not be distinguishable from those of the pure forms. Even if one of these forms were present in only a ratio of 3:1, then the color of this solution will appear to be about the same as that of the predominant form alone. If $[HIn]:[In^-] = 3$ or greater, the solution will be decidedly red. If the ratio of $[In^-]:[HIn]$ is 3 or greater, the solution will be decidedly yellow. The pH of these solutions will be given by

$$pH_{red} \leq pK_{HIn} + \log 1/3 \text{ and } pH_{yellow} \geq pK_{HIn} + \log 3$$

The pH range of perceptible color changes is, then

$$pH = pK_{In} \pm 0.5$$

Although the ratio 3:1 is fairly reasonable for many two-color indicators, the specific value that applies may vary from one indicator to another depending on the $\epsilon$ values of the two conjugate forms. Below are tabulated the characteristics of some popular indicators.

### TABLE 8.1  Selected Acid Base Indicators

| INDICATOR NAME | $pK_{In}$ at $I = 0.1$ & $25°C$ | Acid/Base Colors |
|---|---|---|
| Methyl Orange | 3.46 | Red/Yellow |
| Bromcresol Green | 4.66 | Yellow/Blue |
| Methyl Red | 5.00 | Red/Yellow |
| Bromthymol Blue | 7.10 | Yellow/Blue |
| Phenolphthalein | ~9.0 | Colorless/Red |

To summarize, the **center of the color change** is seen to be $pH = pK_{In}$; changes from one form to the other are visible only in the pH range of $pK_{In} \pm 0.5$. The pH range in which any particular indicator can be used, therefore, depends on its $pK_a$ value. One of the earliest "physical-organic" studies relating molecular structure to chemical behavior was conducted in the 1920s and 1930s by Kolthoff and his students in their desire to develop color indicators applicable to various pH ranges.

As we have already seen, titration curves exhibit a large pH change in the vicinity of the equivalence point. By the addition of a small quantity of properly selected indicator we can observe a color change close to the equivalence point and thereby locate the endpoint. To the extent that the indicator endpoint and the true equivalence point differ, a **titration error** occurs, measured either in titrant volume or, as relative error, in percent.

When dealing with a one-color indicator such as phenolphthalein, i.e., only In is colored, then the appearance or disappearance of color depends solely on the [In] which the eye can detect. (Keep in mind that if we can use a spectrophotometer, then this would be a two-color indicator also. The spectrum of the acid form is in the UV region.) For phenolphthalein this concentration is approximately $5 \times 10^{-6}$ molar. The pH at which this occurs depends on the total concentration of indicator used. Thus, if $C_{In}$ is the total indicator concentration, then

$$[HIn] = C_{In} - [In] \text{ and } K_{In} = [H^+][In]/(C_{In} - [In])$$

the $[H^+]$ at which color appears is

$$[H^+] = K_{In} \{(C_{In}/[In^-]) - 1\}$$

## Example 8.4

At what pH will the pink color of phenolphthalein appear in 100 ml of a solution to which 3 drops of 0.03 M phenolphthalein have been added? $pK_2$ (phenolphthalein) = 9.70.

If we assume 30 drops per milliliter then 0.10 mL have been used.

$$C_{HIn} = 0.10 \times 0.03/100 = 3 \times 10^{-5} \text{ M}$$

$$[H^+] = 10^{-9.70} [3 \times 10^{-5}/5 \times 10^{-6} - 1] = 10^{-9.00}$$

Hence, the pH at which color appears is 9.00. If 30 drops rather than 3 drops of the indicator solution were added, this pH would have dropped to 7.92.

The behavior of two-color indicators contrasts sharply with that of one-color indicators, as we saw above. With the former, the pH of the color transition is essentially independent of total indicator concentration.

As mentioned above, the indicator, itself an acid or base, consumes titrant. Hence, even though only a small amount of indicator is used, a lower limit of the concentration of the substance being titrated is imposed. If extremely dilute solutions must be titrated, then indicators which are especially sensitive must be used, or replaced entirely by a pH meter. The pH meter can be considered to be the **universal indicator**.

## Titration Errors

The error in a titration is the difference between the volume of titrant required to exactly react with an analyte and the volume at which the indicator endpoint is recognized. The use of the spreadsheet-generated titration curve presents us with a uniquely simple and convenient method for describing titration errors. Simply **zoom** in on the equivalence point region of the curve, i.e., after the curve is displayed, change the scale of F (or V) to where F ranges from 0.99 to 1.01 full scale (with V, use 0.99V and 1.01V as limits). From this graph, read the pH values at these limits which correspond to $\pm 1\%$ error. For 0.1% and 0.5% errors, read pH values at F = 0.999 to 1.001 and 0.995 to 1.005, respectively. In addition to this simple, yet empirical, treatment, titration errors will also be discussed under the section on Sharpness Index.

# TITRATION CURVE CALCULUS

Refinement of the use of $\Delta F/\Delta pH$ and its reciprocal as buffer and sharpness indices (as mentioned above) is achieved by obtaining the derivative of the titration curve, $dF/dpH$. As will be seen, this derivative is a function of the several $d\alpha_i/dpH$ values, which in turn are simple functions of the $\alpha$ values themselves.

It is interesting to note that the changes in both $[H^+]$ and $[OH^-]$ with pH are proportional to $[H^+]$ and $[OH^-]$, respectively.

Since

$$\log [H^+] = -pH, \quad d\,[H^+]/2.303\,[H^+]$$

or

$$d[H^+]/dpH = -2.303\,[H^+] \qquad (8\text{-}5)$$

Similarly, as

$$[OH^-] = K_w/[H^+]$$

$$d[OH^-] = -K_w\,d[H^+]/[H^+]^2 = -[OH^-]\,d[H^+]/[H^+]$$

Hence,

$$d[OH^-]/dpH = -2.303\,[OH^-] \qquad (8\text{-}6)$$

For a monoprotic acid,

$$\alpha_o = [H^+]/([H^+] + K)$$

Differentiation gives

$$d\alpha_o/d[H^+] = K/([H^+] + K)^2 = \alpha_o\alpha_1 d[H^+]/[H^+] \qquad (8\text{-}7)$$

Using Equation 8-7, we have

$$d\alpha_o/dpH = -2.303 \; \alpha_o\alpha_1 \qquad (8\text{-}8)$$

Similarly,

$$d\alpha_1/dpH = 2.303 \; \alpha_1\alpha_o \qquad (8\text{-}9)$$

Equations 8-8 and 8-9 can be transformed ($d\alpha_i/2.303\alpha_i = \log\alpha_i$) to express the variation of log $\alpha$ with pH

$$d \log \alpha_o/dpH = -\alpha_1$$

$$d \log \alpha_1/dpH = -\alpha_o \qquad (8\text{-}10)$$

Thus, at high pH ($>>pK$), when $\alpha_1 = 1$ ($\alpha_o \cong 0$), the slope of the log $\alpha_o$ vs. pH will be -1. Conversely, at low pH, the slope of the log $\alpha_o$ curve is zero and that of log $\alpha_1$ is +1.

For diprotic acids the derivatives of $\alpha$ values with respect to pH can be shown to be

$$d\alpha_o/dpH = 2.303 \; \alpha_o[-\alpha_1 - 2\alpha_2]$$

$$d\alpha_1/dpH = 2.303 \; \alpha_1[\alpha_o - \alpha_2]$$

$$d\alpha_2/dpH = 2.303 \; \alpha_2[\alpha_1 + 2\alpha_o] \qquad (8\text{-}11)$$

In each of these cases, the $d\alpha/dpH$ is proportional to the $\alpha$ in question times a factor which includes all the other $\alpha$ values, each multiplied by the number of protons gained (a minus sign for protons lost). The factor containing the other $\alpha$ values represents the slope of the log $\alpha$ - pH curve, i.e., the EquiligrapH. Thus, the slope of the log $[CO_3^{2-}]$ vs. pH curve will be +2 when $H_2CO_3$ predominates ($\alpha_o \approx 1$, $\alpha_1$ and $\alpha_2$ being almost zero),

+1 when $HCO_3^-$ predominates ($\alpha_1 \approx 1$, the other $\alpha$s $\approx 0$) and 0 when $CO_3^{2-}$ predominates ($\alpha_2 \approx 1$, the other $\alpha$s $\approx 0$).

We can generalize from these considerations that one can write the value for $d\alpha/dpH$ for any $\alpha$ in an N-protic acid system as 2.303 times the product of two factors (1) the $\alpha$ in question and (2) the sum of all the other alphas in the system and the number of protons gained in going from the species corresponding to the $\alpha$ in question to that of each of the other $\alpha$s. Thus, to illustrate with $d\alpha_2/dpH$ for $H_4Y$ (EDTA)

$$d\alpha_2/dpH = 2.303 \; \alpha_2[-\alpha_3 - 2\alpha_4 + \alpha_1 + 2\alpha_0] \qquad (8\text{-}12)$$

With this brief application of differential calculus, let us return to questions relating to the sharpness of titration endpoints, buffer indices, and similar characteristics of acid-base mixtures. In effect, the slope of the function of Equation 8-3, $dF/dpH$, is a measure of the buffer capacity and its reciprocal, $dpH/dF$, evaluated at the equivalence points, measures the sharpness index of the corresponding titration.

It is convenient to consider F as consisting of two parts, one solely a function of $\alpha$ values, the second a function of concentration $C_A$, $[H^+]$, and $[OH^-]$ of the point under consideration. Differentiation of the first part, $\sum i\alpha_i$, with respect to pH will of course depend on N, the number of dissociable protons. Thus, for a monoprotic acid,

$$\frac{d}{dpH}\sum_0^N i\alpha_1 = 2.303\alpha_0\alpha_1 T \qquad (8\text{-}13)$$

and for a diprotic acid,

$$= 2.303 \; \{\alpha_0(\alpha_1 + 4\alpha_2) + \alpha_1\alpha_2\} \qquad (8\text{-}14)$$

for a triprotic acid,

$$= 2.303 \; \{\alpha_0(\alpha_1 + 4\alpha_2 + 9\alpha_3) + \alpha_1(\alpha_2 + 4\alpha_3) + \alpha_2\alpha_3\} \qquad (8\text{-}15)$$

and, for a tetraprotic acid,

$$= 2.303 \; \{\alpha_0(\alpha_1 + 4\alpha_2 + 9\alpha_3 + 16\alpha_4) + \alpha_1(\alpha_2 + 4\alpha_3 + 9\alpha_4) + \alpha_2(\alpha_3 + 4\alpha_4) + \alpha_3\alpha_4\} \qquad (8\text{-}16)$$

In these expressions, it can be seen that, providing there is a reasonable separation of successive pK values, only the first term in each of the factors in parentheses is of significance. Thus, for such a triprotic acid

$$\frac{d}{dpH}\sum_0^N i\alpha_i - 2.303(\alpha_0\alpha_1 + \alpha_1\alpha_2 + \alpha_2\alpha_3) \qquad (8\text{-}17)$$

The differential of the other part of F,

$$d/dpH \{(V_A + V_B) C_A([OH^-] - [H^+])\}/V_A C_A$$

is

$$= 2.303(V_A + V_B) ([OH^-] + [H^+]) V_A C_A \qquad (8\text{-}18)$$

which naturally retains the same form in all cases.

## Buffer Index

The buffer index is a measure of the amount of acid or base that can be added to a solution before the pH changed by a given amount. If Equations 8-17 and 8-18 are rewritten as

$$C_B = C'_A F = C'_A \sum i\alpha_i - [H^+] + [OH^-] \qquad (8\text{-}19)$$

Where

$$C'_A = C_A [V_A/(V_A + V_B)] F$$

Then

$$dC_B/dpH \equiv \text{Buffer index} = C'_A \, dF/dpH \qquad (8\text{-}20)$$

For a monoprotic acid, therefore, the buffer index is

$$dC_B/dpH = 2.303 \{C'_A \, \alpha_0\alpha_1 + [H^+] + [OH^-]\} \qquad (8\text{-}21)$$

For a diprotic acid, it is

$$= 2.303 \{C'_A(\alpha_0\alpha_1 + \alpha_1\alpha_2) + [H^+] + [OH^-]\} \qquad (8\text{-}22)$$

and that for a triprotic acid is

$$= 2.303 \{C'_A(\alpha_o\alpha_1 + \alpha_1\alpha_2 + \alpha_2\alpha_3) + [H^+] + [OH^-]\} \qquad (8\text{-}23)$$

## Sharpness Index and Titration Errors

The inverse of the slope of the titration curve is a measure of how "sharp" the curve is. Thus, if $V_{titrant}$ or F changed by 1% in the vicinity of the endpoint cause a pH change of 1 unit, the **sharpness index**, or **dpH/dF** would be 1/0.01 or 100. This is a measure of the feasibility of conducting a titration within a given limit of error. This is closely related to the **titration error**, which measures the error in absolute, i.e., titrant volume, units, or relative units, i.e., $\Delta F$. Earlier, an empirical method of obtaining titration errors based on using spreadsheet graphs, was described. In this section, equations for titration error will be derived and results from each will be compared.

A few examples will serve to illustrate the usefulness of these concepts. First, what factors determine the titration feasibility of a monoprotic acid?

For the monoprotic acid case, using $C'$ as the concentration corrected for dilution, we have

$$dF/dpH = 2.303 \; \alpha_o\alpha_1 + 2.303([H^+] + [OH^-])/C'_A \qquad (8\text{-}24)$$

In the vicinity of the endpoint, $\alpha_1 \approx 1$, $\alpha_o = [H^+]/K$ so that

$$dF/dpH = 2.303 \{[H^+] \; (1/K + 1/C'_A) + [OH^-]/C'_A\} \qquad (8\text{-}25)$$

Usually at the endpoint $[OH^-] >> [H^+]$ and from the approximate solution of the pH at the equivalence point

$$[OH^-] \approx (C'_A K_w/K)^{\frac{1}{2}} \qquad (8\text{-}26)$$

giving

$$dF/dpH = 2.303 \; (K_w/C'_A K)^{\frac{1}{2}} \qquad (8\text{-}27)$$

and finally, the sharpness index, S.I.

$$S. \; I. = dpH/dF = 0.43 \; (C'_A K/K_w)^{\frac{1}{2}} \qquad (8\text{-}28)$$

For a titration of acetic acid (pK = 4.74) using equimolar NaOH (so that $C'_A = C_A/2$)

$$S. I. = 10^{-0.52} (10^{-4.74}/10^{-14.00})^{1/2} \cdot C'^{1/2}_A = 10^{4.11} C''^{1/2}_A \qquad (8-29)$$

The sharpness index of the HOAc-NaOH titration $\leq 10^3$ when the initial concentration of HOAc is 0.013 M or lower. In general, the sharpness of a titration of a weak monoprotic acid varies with the **square root** of its **acid dissociation constant** and its **initial concentration**.

Equation 8-28 is formally analogous to what would be obtained at the second equivalence point of a diprotic acid titration or the last equivalence point of an N-protic acid titration. In these cases, the appropriate acid dissociation constant, $K_2$ or $K_N$, must of course be used.

In the vicinity of the equivalence point, when the change in F of interest is small, say under 0.05, Equation 8-27 may be readily revised to give an expression for the **titration error**, $\Delta F$, as a function of pH change, $\Delta pH$. The titration error is proportional to the reciprocal of the S. I.

$$\Delta F = 2.303(K_w/C'_A K/)^{1/2}\Delta pH$$

Compare the results predicted from this equation and those described below for other titrations with those obtained by a graphical solution (using a greatly expanded F scale from 0.99 to 1.01 and reading the pH from the graph).

Now let us turn to the factors that determine the sharpness at an intermediate equivalence point such as that of a half-neutralized diprotic acid.

From Equations 8-17 and 8-18 we have

$$dF/2.303dpH = \alpha_o\alpha_1 + \alpha_1\alpha_2 + \{[H^+] + [OH^-]\}/C'_A \qquad (8-30)$$

In the vicinity of the first endpoint:

$$\alpha_1 \approx 1, \ \alpha_o \approx [H^+]/K_1, \text{ and } \alpha_2 \approx K_2/[H^+]$$

Hence,

$$dF/2.303dpH \approx \alpha_o + \alpha_2 + \{[H^+] + K_w/[H^+]\}/C'_A \qquad (8-31)$$

or

$$\frac{dF}{2.303pH} - \frac{[H^+]}{K_1} + \frac{K_2}{[H^+]} + \frac{[H^+]}{C_A} + \frac{K_W}{[H^+]C_A}$$

$$- \frac{C_A([H^+]^2 + K_1K_2) + K_1[H^+]^2 + K_W)}{K_1[H^+]C_A}$$

(8-32)

Because at this equivalence point, $[H^+]^2 \approx K_1K_2$, we now have

$$dF/2.303dpH \approx \{2C'_A(K_1K_2) + K_1(K_1K_2 + K_w)\}/K_1(K_1K_2)^{1/2}C'_A$$  (8-33)

Inasmuch as in any practical titration

$$2 K_1K_2C'_A >> K_1(K_1K_2 + K_w)$$  (8-34)

or

$$2 C'_A >> K_1 + K_w/K_2$$  (8-35)

$$dF/2.303dpH \approx 2(K_2/K_1)^{1/2}$$

The **titration error** is

$$\Delta F = 4.606(K_2/K_1)^{1/2}\Delta pH$$

and the sharpness index is

$$S.I. = dpH/dF \approx 1/4.6 (K_1/K_2)^{1/2}$$  (8-36)

which we see to be **independent of initial concentration** of the titrant and proportional to the ratio of the first and second acid dissociation constants. The sharpness index for carbonic acid ($pK_1 = 6.13$ and $pK_2 = 10.33$) titrated to the first equivalence point is only about 25. This would seem to limit such a titration to a minimum error of 5% or, if the pH of the equivalence point could be matched to $\pm 0.5$, to 2.5%.

In addition to being very helpful in the detailed analysis of acid-base titrations, the treatment outlined here may be readily adapted to complexometric titrations (Chapter 9).

## Locating the Equivalence Point

We have already discussed two methods of locating the equivalence points in titrations: (1) as the center of the vertical portion (i.e., the most rapid change of pH with titrant volume) of the curve and (2) the steepest point in the plot of titration curve slope ($\Delta pH/\Delta V$ vs. pH). Both of these methods require that very precise and closely spaced data be taken in the vicinity of the equivalence point.

It is possible to significantly improve the reliability as well as simplicity of locating the equivalence point by converting the logarithmic expression of critical variable concentration, i.e., the pH, to a linear one, $[H^+]$. This suggestion was first advanced in 1952 by Gunnar Gran, a Swedish analytical chemist but could not be exploited for several decades because commercial pH meters were not capable of sufficiently reliable pH measurements.

## Gran Method

The **Gran Method** is particularly valuable because, like most based on linear titrations, it yields straight line segments susceptible to least squares method analysis, which leads to great accuracy in equivalence point determination.

Let us consider a typical point in the titration of $V_A$ mL of a strong acid of unknown concentration $C_A$, with $V_B$ mL of a known concentration, $C_B$ M of NaOH. The pH of this solution is converted to the corresponding $[H^+]$ ($= 10^{-pH}$). Then the number of millimoles of $H^+$ not yet neutralized, can be expressed as either $[H^+]$ ($V_A + V_B$) or $C_A$ ($V_E - V_B$), where $V_E$ is the volume of base at the equivalence point. At every point before the equivalence point,

$$[H^+] (V_A + V_B) = C_A (V_E - V_B) \qquad (8\text{-}37)$$

A plot of $[H^+]$ ($V_A + V_B$) vs. $V_B$ will give a straight line intersecting the V axis at $V_B = V_E$. Two important advantages result: (1) the values of $V_B$ used in the titration do not have to be close to the equivalence point. Time is saved when it is not necessary to make careful, drop-wise additions in the vicinity of the equivalence point and   (2) the line can be most accurately defined by use of the linear regression command /**TAR**{/AR;/DAR}.  The increase on precision of $V_e$ is a direct consequence of the utilization of all of the titration data, not merely the few points taken close to the equivalence point, as is the case in the conventional titration.

Similarly, one can use the data obtained after the equivalence by equating the two ways of expressing the excess base.

$$(V_A+V_B)\ [OH^-] = C_B(V_B-V_E); \text{ since } [OH^-] = K_w/[H^+],$$

$$(V_A+V_B)10^{pH} = C_B(V_B-V_E)/K_w \qquad (8\text{-}38)$$

A plot of the left-hand side of Equation 8-40 vs. $V_B$ will also intersect the V axis at $V_B = V_E$.  Moreover, the value of $K_w$ can be obtained from the slope of this linear plot.

For the titration of a weak monoprotic acid with NaOH, the linear relation describing points before the equivalence point is

$$V_B 10^{-pH} = K_a(V_E - V_B) \qquad (8\text{-}39)$$

which may be seen as a version of the simplest equation for a buffer $V_E$ $V_B$ is the concentration of weak acid remaining after addition of $V_B$ mL of NaOH which in turn has totally reacted to form the conjugate base of the weak acid).

A plot of $V_B 10^{pH}$ vs. $V_B$ not only locates $V_E$ as the intersection of the line with the V axis, but its slope yields the value of $K_a$.  After the equivalence point, the appropriate equation is the same as Equation 8-41, inasmuch as the conjugate base of a weak acid is not active in the presence of excess NaOH.

In both strong/weak acid titrations with NaOH then, the Gran relationships lead to give a **LINEAR** titration curve that permits high precision in locating the equivalence points, particularly in problem cases where the acid is very weak or its concentration is very low.  Further, determination of the slope and intercepts of the linear portions of the curves can be readily accomplished by the linear regression command /**TAR**{/AR; /DAR}.

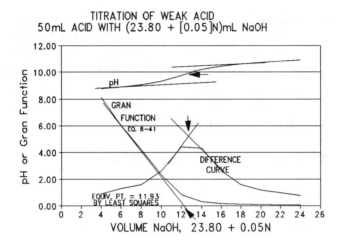

**Figure 8.6   Locating the Equivalence Point by Difference Titration and Gran Methods**

A comparison of the three methods of locating the equivalence point described above is illustrated in Figure 8.6 where a rather weak acid is titrated with NaOH. When the volume of base was within ca. 1 mL of the equivalence point, further pH data was obtained by drop- wise (0.05 mL) addition of NaOH. The value of $V_E$ can be obtained by

(1) visual location of the midpoint of the steepest portion of the pH change, (2) finding the intersections of the linear portions of the different curves, or (3) finding the intersection of the line derived from the points before the equivalence point. As may be seen from Figure 8.6, these methods locate the equivalence point to within less than 0.5 drops or $\pm0.02$ mL. The most precise location of $V_E$ is obtained by using linear regression to define the best line using the points taken well before $V_e$. From the slope of this line, $K_a$ was estimated as $9.7 \times 10^{-8}$.

## PROBLEMS

[Note for these problems and all others with multicurve comparison graphs: Enter the values of $C_A$, $V_A$, $C_B$, pK, and $pK_2$ either in absolute cell address or Block Name style in the spreadsheet formulas. In this way, one spreadsheet will serve for all five parts of this question. Further, before changing parametric values, which will bring changes in the titration volume column, save the values in a conveniently located column using the Values command(/EV{/BAV; /RV).]

8-1     Develop five spreadsheets for drawing the titration curves for

(a)            50 mL of 0.05 M HOAc with 0.10 M NaOH

(b)            25 mL of 0.20 M $NH_3$ with 0.10 M HCl

(c)            25 mL of 0.10 M ethylenediamine with 0.10 M HCl

(d)            50 mL of a mixture of 0.10 M HOAc and 0.10 M $NH_4Cl$
               with 0.10 M NaOH

(e)            50 mL of 0.10 M of a diprotic acid, $H_2B$, with 0.10 M
               NaOH when the pK values of $H_2B$ are (i) 4 and 8, (ii) 3
               and 8, (iii) 2 and 8, (iv) 4 and 7, (v) 4 and 6. What can
               you conclude from a comparison of the shapes of these
               curves?

(f)            Calculate discrete points for titrations in (a), (b), and (c)
               at 0%, 30%, 100%, and 150% neutralization.

8-2     For the systems described in 10.1, draw sharpness and buffer indices.

8-3     How far apart must successive $pK_a$ values be in order to have clearly
        distinguishable endpoints for titration curves of polyprotic acids?
        Consider as examples $H_3PO_4$ and citric acid. What about equivalence
        points at very high or low pH? How easy are they to locate?

8-4     How does the maximum in the $\triangle pH/\triangle V_B$ illuminate the nature of
        the titration? For one, it locates the endpoint precisely. Also, the
        amplitude is a measure of the sharpness index of the titration. Show
        that this $(\triangle pH/\triangle V_B)_{max}$ is proportional to $C_A^{\frac{1}{2}}$.

8-5     With points on the titration curve closely spaced, obtaining slopes ($\Delta pH/\Delta V_B$) is a numerical differentiation that is a reasonable approximation of the calculus operation of differentiation. Compare the numerical and calculus approaches in titration curves using several different acid concentrations.

8-6     For the titrations done in Problem 8.1, calculate titration errors by the empirical (graphical; see above, "Characterizing Titration Curves") and $\Delta F$ equations and compare the results.

8-7     Compare the Gran titration and the derivative titration methods for precise location of equivalent point for the system $H_3PO_4$ NaOH. First generate the data set for the titration curve then add "noise" (see Chapter 12) to both the titration volume and pH values using +0.02 (@RAND-@RAND) to add to the volumes and +0.01 (@RAND-@RAND) to pH values.

8-8     Compare the titration curve obtained in Figure 8.3 with the one obtained in the presence of 0.15M $Ca^{2+}$ (Remember complex formation: $Ca + HPO4) = Ca(PO_4) + H^+$).

# Chapter 9

# Titrations II
# Lewis Acids and Bases

## COMPLEXOMETRY

The general requirements for useful titrations described in Chapter 8, certainly apply to those involving metal ions and ligands, e.g., that the titration reaction must go to completion in a relatively rapid fashion and be free of any interfering side reactions. In order for the reaction to go to completion, the equilibrium constant governing it must be fairly large. Such is the case with most acids reacting with NaOH, and bases reacting with HCl. Furthermore, the self-dissociation constant of the solvent water provides a large enough difference between the pH at the equivalence point and that in a slight excess of strong base (or acid) that the endpoint

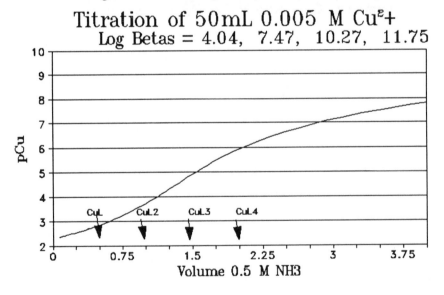

Figure 9.1 Titration curve of $Cu^{2+}$ with Ammonia

178

location is reasonably reliable in most titrations.

In contrast, in metal ligand titrations, the self-dissociation of water is not a directly relevant reaction. The rise in the pM value after the equivalence point is a function of the log $\beta$ values(s) of the complex, which therefore must be larger than the $pK_a$ values that can be tolerated in pH titrations.

Another difference lies in the polyfunctional nature of metal ions. They are capable of reacting with more than one ligand, forming a series of complexes whose equilibrium constants are much more closely spaced than those of most polyprotic acids. For example (Figure 9.1), the titration of $Cu^{2+}$ with $NH_3$ would certainly be of dubious practical analytical value. Copper(II) ion forms a series of ammonia complexes whose formation constants are both small and closely spaced; there are several complex species present in significant concentrations throughout the titration.

A better titration can be achieved using a bidentate ligand, like ethylene-diamine, whose formation constants are $10^{10.48}$ and $10^{8.34}$ (Figure 9.2).

The most efficient titration will occur when the complexation takes place in a single step, so that the vertical portion of the curve reflects the generally higher formation constant observed (Figure 9.3).

In this chapter, equations will be developed for complexometric titration curves along similar lines used in the previous chapter for Brønsted acid-

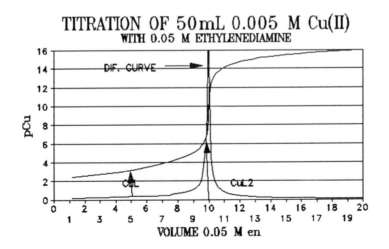

**Figure 9.2 Titration of $Cu^{2+}$ with Ethylenediamine**

base titration curves. Various means of locating the endpoints will be described, and metallochromic indicators will be described.

In Lewis acid/base titrations, the concentration variable comparable to $[H^+]$ is the ligand concentration, but the volume of ligand or the titration fraction is plotted against pM or $[M^{n+}]$, depending on whether the titration is being plotted logarithmically or linearly. This is no cause for concern since we will be able to calculate this quantity, like all the others we need, from the ligand concentration.

## Derivation of Complexometric Titration Curve Equations

A general titration curve equation can be derived from the mass balance equation for the ligand, i.e., to state that the total ligand concentration is the sum of the bound and free ligand concentrations. By the bound ligand is meant the ligand incorporated into the metal complex(es). Thus, for $Ag^+$ and $NH_3$,

$$\text{Bound Ligand} = [Ag(NH_3)^+] + 2[Ag(NH_3)_2^+]$$

Again, for $Fe^{3+} + Cl^-$;

$$\text{Bound Ligand} = [FeCl^{2+}] + 2[FeCl_2^+] + 3[FeCl_3] + [FeCl_4^-]$$

The concentration of each of these species is given by the $\alpha C$ product which, in turn, depends on the value of a single variable, $[L]$. To express the relationship that total ligand concentration, $C_L$, is comprised of bound and free ligand concentrations, we can write

$$\frac{V_L C_L}{V_L + V_M} - \frac{V_M C_M}{V_L + V_M} (\alpha_{ML} + 2\alpha_{ML_2} + \ldots n\,\alpha_{ML_n}) + [L^-] \qquad (9\text{-}1)$$

The $\alpha$ factor $(\alpha_{ML} + 2\alpha_{ML2} + n\alpha_{MLn})$ is also called $\bar{n}$, the formation function (notice its equivalent expression: $(N - \sum^N_i i\alpha_i)$ in Equation 8-7). Hence, Equation 9-1 can be written as

$$\frac{V_L C_L}{V_L + V_M} - \frac{V_M C_M}{V_L + V_M}\,\bar{n} + [L] \qquad (9\text{-}2)$$

or, rearranging,

$$V_L C_L = V_M(C_M \bar{n}) + [L](V_M + V_L)$$

and, finally,

$$V_L - V_M \frac{(C_M \bar{n} + [L^-])}{(C_L - [L^-])}$$

*or* 

(9-3)

$$F \equiv \frac{C_L V_L}{C_M V_M} - \bar{n} + \frac{V_L + V_M}{C_M V_M} \cdot [L]$$

It is useful to compare Equation 9-3 with Equations 8-6 and 8-7. An interesting analogy links the base, which is capable of binding a number of protons, to the metal ion, which is capable of coordinating a number of ligands. Comparing $\bar{n} C_B$ to $\bar{n} C_M$ and $[H^+]$ to $[L]$, the only difference between Equations 8-6 and 9-3 is the absence of $[OH^-]$. In metal complexation, water is not a direct participant; hence, there is no $[OH^-]$.

The titration curves in Figures 9.1 and 9.2 are derived from Equation 9-3. The spreadsheet is organized by using /EF{/BF; /DF} in column A for a series of log[L] values ranging from 2 units less than $-\log\beta_1$ to a value of within 1 unit of log $C_L$, in steps of 0.1 or 0.2 increments. In successive columns, place values of $[L]$, $\alpha_M$, $\alpha_{ML}$, $\alpha_{ML2}$,...$\alpha_{MLn}$, and, with the help of Equation 9-3, $V_L$. Finally, a column of $[M]$ values, which is $V_M C_M \alpha_M / (V_M + V_L)$. Values for $V_M$ and $C_M$ will be determined by the conditions cited in the problem.

## Example 9.1

Consider an idealized titration (no metal complexation or pH effects on ligand or metal) of 50 mL of 0.003 M $Zn^{2+}$ with 0.01 M $Na_2H_2Y$(EDTA) and draw the titration curve. Develop the spreadsheet by block filling **A** with log$[Y^{4+}]$ values from -18 to -3 in 0.3 increments. In successive columns, place values of $[Y_{4-}]$, $\alpha_{Zn}$, $\alpha_{ZnY}$, and with the help of Equation 9-3 $V_Y$. Finally, a column of $[Zn]$ values, which is $V_{Zn} C_{Zn} \alpha_{Zn} / (V_{Zn} + V_Y)$.

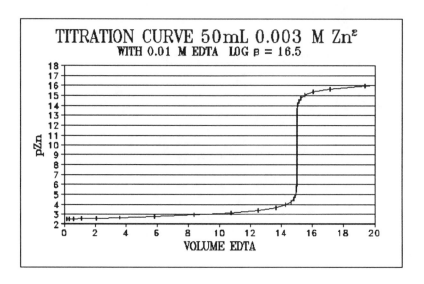

**Figure 9.3 Idealized Titration of 0.003 M Zn²⁺ with 0.01 M EDTA**

Titrations of metal ions with ligands as titrants are far more frequently encountered than those in which ligand solutions are titrated with metal ions as titrants. These are of practical utility, nevertheless, as means of determining compounds which happen to be ligands of reasonably strong complexing power. The titration curve Equation 9-3 can be readily adapted to give Equation 9-4. Figure 9.3a shows a titration curve obtained with Equation 9-4 from the titration of 50 mL of a 0.05 M ligand which forms two $Cu^{2+}$ complexes (log $\beta_1$ = 11, log $\beta_2$ = 17) with 0.05 M $Cu^{2+}$ as titrant.

$$V_M = \frac{(V_L - [L])}{(C_M \bar{n} + [L])} \tag{9-4}$$

## Role of pH and Auxiliary Ligands in Complexometric Titrations

The titration curve developed in Example 9.1 was hypothetical. In practice, complexometric titrations are conducted at pH values at which the titrant is protonated to some extent. Often, an auxiliary complexing agent such as ammonia is present so that there are various metal-containing species in addition to just the simple hydrated metal ions. Calculations of such titration curves can be readily accomplished within the framework of the material presented in Chapter 5, particularly that dealing with $\beta'$, the conditional constant. In the treatment that follows, side reactions of the titration ligand and metal ion will be considered using the familiar $\alpha_L$ and $\alpha_M$ terms. Because there are relatively few examples of side reactions involving the titration complex, they will be neglected for the present.

### Example 9.2

What changes are needed to modify the approach taken in Example 9.1 to deal with a Zn-EDTA titration if it is conducted in the presence of 0.1 M $NH_3$ and at a pH of 9.00?

$$\alpha_Y = 10^{-10.33}/(10^{-9.00} + 10^{-10.33}) = 10^{-1.33} \tag{9-5}$$

$$\alpha_{NH3} = 10^{-9.27}/(10^{-9.00} + 10^{-9.27}) = 10^{-0.46}$$
$$[NH_3] = \alpha C = 10^{-1.46} \tag{9-6}$$

$$\alpha Z_n = 1/(1 + 10^{2.32}[NH_3] + 10^{4.81}[NH_3]^2 + 10^{7.11}[NH_3]^3 + 10^{9.32}) = 10^{-3.56}$$

Therefore, $\beta'_{ZnY}$ is

$$\beta'_{ZnY} - \beta_{ZnY} \, \alpha_Y \alpha_{Zn} - \frac{[ZnY]}{C_{Zn} \, C_Y} \tag{9-7}$$

Equation 9-3, may be used as before, except that we substitute $\beta'$ for $\beta$ and $C_{Zn}$ for $[Zn^{2+}]$. This spreadsheet can be organized exactly as those for Figures 9.1, 9.2, and 9.3. It is noteworthy that while pZn starts higher when $NH_3$ is present, the two curves of 9.3 and 9.5 would coincide after the equivalence point, where EDTA is in excess. These two figures appear to contradict this only because no $\alpha_Y$ correction was carried out for the curve in 9.3.

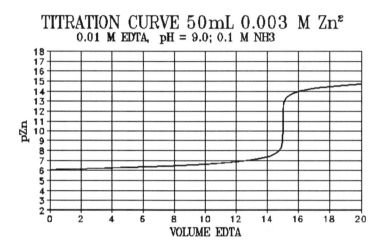

**Figure 9.3a Titration of Zn-EDTA at pH9.0 with $NH_3$**

### Calculations of pM for Selected Individual Points

As shown in Chapter 8, selected individual points in complexometric titration curves can be calculated without recourse to the titration curve equation by (a) performing the reaction stoichiometry at the desired point, (b) recognizing the type of equilibrium calculation represented by the mixture, and (c) using the appropriate formula to find pM. Thus, with a Cu-EDTA titration in the presence of ammonia, all points before (and not very close to) the equivalence point, $[Cu^{2+}] = \alpha_{Cu}C'_{Cu}$, where $C'_{Cu}$ is $(V_{Cu}C_{Cu} - V_YC_Y)/(V_{Cu} + V_Y)$ and $\alpha_{Cu}$ has its usual meaning. For any point beyond the equivalence point, the metal-complex analog to the buffer equation gives

$$pCu = \log(\beta'_{CuY}/\alpha_{Cu}) + \log)C'Y/[CuY])$$

where C'Y represents the concentration of the unreacted EDTA.

## Metallochromic Indicators

The development of visual indicators for use in complexometric titrations even in very dilute solutions has provided a stimulus for the increasing popularity of titrimetric determinations of metal ions. Such indicators behave in an exactly analogous manner as the acid-base indicators.

**TABLE 9.1  SELECTED METALLOCHROMIC INDICATORS**

| Name | $pK_a$ Values and Colors | Log $\beta_{MIn}$ and Colors |
|---|---|---|
| Erio Black T (E-BT) ($H_2In^-$) | 6.3, 11.6 red→blue→orange | 12.9(Zn), 9.6(Mn), 7.0(Mg), 5.4(Ca) MIn red |
| Calmagite ($H_2In^-$) | 8.1,12.1 red→blue→orange | 8.1(Mg), 6.1(Ca) MIn red |
| Pyrocatechol Violet($H_3In$) | 7.8, 9.8, 11.7 red→yel→viol→ red-purple | 27.1(Bi), 23.1(Th) MIn |
| Xylenol Orange ($H_5In$) | 2.6, 3.2, 6.4, 10.5, 12.3 yel≤ pH6.4 ≥violet | log $\beta'$(pH 6)=7.4(Hg),7.0(Pb), 4.5(Cd,Zn,La)          MIn blue |

A metallochromic indicator is a substance capable of giving a highly colored, water-soluble complex whose stability is such that is log $\beta'_f$ is close to the value of pM at the equivalence point of the titration. Most metallochromic indicators are colored, weak Brønsted acids having acid-base indicator properties as well as metal ion indicator properties so that color changes will depend on pH. This is one of the reasons for maintaining pH control through the use of buffers during a complexometric titration. In some instances, the indicator is colorless, e.g., sulfosalicylic acid, in analogy with one-color acid-base indicators.

One of the widely used indicators for calcium is Eriochrome Black T which is a sulfonated, o,o′-dihydroxy azo dye that can be represented by the formula $H_2In^-$. The dissociation of the two protons can be represented by the equations:

$$H_2In^- \rightleftharpoons HIn^= + H^+ \ (pK_1 = 6.3)$$
$$\text{winered} \quad \text{blue}$$

$$HIn^= \rightleftharpoons In^- + H^+ \ (pK_2 = 11.5)$$
$$\text{blue} \quad \text{orange}$$

Most metal ions form intensely colored, red complexes with Eriochrome Black T. However, a number of these are so stable that the metal is not readily displaced from the dye by the addition of small amount of chelating titrant beyond the end-point. ($\log \beta_{MIn} >> pM$ at the equivalence point.) In other instances, although the stabilities are appropriate, the rates of transformation of the metal complexes are too slow.

The choice of conditions for the use of Eriochrome Black T as an indicator for the titration, for example, of magnesium with EDTA may be made on the basis of the following considerations. The $\beta'$ values of both the Mg-EDTA and Mg-Indicator vary with pH. Because of the colors of the various indicator forms and their equilibrium constants, the minimum pH at which magnesium may be titrated with EDTA is 10.

We must also keep in mind that in order to avoid having the uncomplexed indicator in the orange form, the pH should be kept sufficiently below 11.5. (The color change from red to blue is more easily seen than that from red to orange.) Hence, for all practical purposes, the pH must be adjusted between 10 and 11. Within this range, the optimum pH would be one in which the $\log \beta'_{MIn}$ is closest to pM at the equivalence point. In the titration of $10^{-2}$ M magnesium, pMg at the equivalence point is given by

$$pMg = 1/2(\log \beta'_{MgY} - \log C_{Mg})$$

This expression is formally analogous to the expression

$$pH = 1/2(pK_a - \log C_{HA})$$

for a weak Brønsted acid, and has been used above for the calculation of pM at the equivalence points of the titration curves of $Ca^{2+}$ and $Cu^{2+}$ vs. EDTA. At pH 10, pMg at the equivalence point is equal to 5.12, since $\log \beta_{MgY} = 8.69$, $\log \alpha_4 = -0.46$ and $\log C_{Mg} = -2.0$.

At the minimum pH = 10.00, $\log \beta'_{MgIn} = \log \beta_{MgIn} + \log \alpha_2$

Since $K_2 = 10^{-11.50}$, $\alpha_2 = K_2/(K_2 + [H^+]) = 10^{-1.51}$

Since $\beta_{MgIn} = 10^{6.95}$, $\beta'_{MgIn} = 10^{5.44}$

In order to reduce the value of log $\beta K'_{MgIn}$ to where it would equal pMg at the equivalence point (i.e., 5.12) it would be necessary to reduce the pH below the minimum value of 10. Hence in this titration the optimum pH is 10. Since the difference between pMg at the equivalence point and log $\beta'_{MgIn}$ is small, the theoretical titration error is small.

In the titration of calcium with EDTA, Eriochrome Black T is not a good indicator unless some magnesium is present. (Usually a small amount of a magnesium salt is added to the EDTA titrant prior to standardization). In the titration of 0.01 M $Ca^{2+}$ with EDTA, in the absence of $Mg^{2+}$, the value of pCa at the equivalence point varies between 5.2 at pH 8 and 6.3 at pH 11. In this pH range, the value of the logarithm of the apparent indicator constant (log $\beta'_{CaIn}$) varies between 2.9 and 4.7. Since log $\beta'_{CaIn}$ is much smaller than pCa at the equivalence point, the color change occurs much too soon and is too gradual to be of use. Raising the pH above 11 will increase $\beta'_{CaIn}$ and thus bring it closer to pCa; this is not practical, however, because the free indicator color changes from blue to orange at pH values higher than 11.

The presence of $Mg^{2+}$ (of Mg-EDTA which is transformed by $Ca^{2+}$ to $Mg^{2+}$) renders Eriochrome Black T a useful indicator for the titration of $Ca^{2+}$ at pH 10, since the indicator complex which it forms is not dissociated until after all of the $Ca^{2+}$ is titrated. At pH 10, pCa at the equivalence point is 6.1. The value of log $\beta'_{MgIn} = 5.4$ is much closer to this pCa than the corresponding value of log $\beta'_{CaIn} = 3.8$.

The use of Eriochrome Black T as an indicator in the $Zn^{2+}$-EDTA titration illustrates a case in which the indicator metal-complex is so stable that the color change occurs after the equivalence point, (log $\beta'_{ZnIn} >$ pZn at equivalence point). Since log $\beta'_{ZnIn}$ changes more rapidly with pH than does pZn (Why?), it is possible to adjust the pH to reduce the difference between them to a reasonably small value (Table 9.1).

**Table 9.1 Titration Characteristics of 0.01 M $Zn^{++}$ with EDTA Using Eriochrome Black T as Indicator**

| pH | pZn at Equivalence Point | log $\beta'_{ZnIn}$ |
|----|--------------------------|---------------------|
| 9  | 8.6                      | 10.4                |
| 8  | 8.1                      | 9.4                 |
| 7  | 7.6                      | 8.4                 |

A practical lower limit is reached at pH 7 since at values lower than this pH the free indicator changes color to red.

It has been tacitly assumed in this section that the maximum color change will occur at the point when free and metal complexed indicator concentrations are equal, i.e., $pM = \log \beta'_{MIn}$. This is only an approximation since the color intensities of these two forms are not necessarily the same.

## Titration Errors

In the last chapter, a graphical method for evaluating titration errors directly from the spreadsheet-generated titration curve was described. Simply **zoom** in on the equivalence point region of the curve, i.e., after the curve is displayed, change the scale of F (or V) to where F ranges from 0.99 to 1.01 full scale (with V, use 0.99 V and 1.01 V as limits). From this graph, read the pH values at these limits which correspond to $\pm 1\%$ error. For 0.1% and 0.5% errors, read pM values at F = 0.999 to 1.001 and 0.995 to 1.005, respectively.

## Locating the Endpoint

Examination of the titration curve visually can locate the endpoint as the midpoint of the nearly vertical portion in the vicinity of the equivalence point. Additionally, if the difference curve in which $\Delta pM/\Delta V_L$ is plotted against $V_L$, is constructed, the maximum is located at the endpoint.

As we found with acid-base titrations in the last chapter, however, the most precise method for locating the endpoint is the **Gran** method. This method results in a linear plot which intercepts the equivalence point at the X axis. Not only can we easily find the best line through linear regression but, as mentioned earlier, the necessary points can be taken at a distance from the equivalence point making this method rapid and convenient.

Assuming that the metal and ligand concentrations are equal, the equation that applies to all points before the equivalence point is

$$10^{-pM}(V_M + V_L) = C_M(V_E - V_L\}$$

With this equation, as with its analog Equation 8-39, a plot of

$10^{pM}(V_M + V_L)$ vs. $V_L$ is linear and crosses the X- axis at $V_L = V_E$.

## Titration Curve Calculus

Complexometric titration curves can be better characterized when we can describe slopes of curves. The change of pM (= -npL) with $V_L$ and its reciprocal will help us understand metal buffer behavior and sharpness of titrations, just as they did with acid-base titrations. The mathematical treatment that follows is abbreviated because of the strong resemblance it bears to that described in the previous chapter.

The derivatives of $\alpha_M, \alpha_{ML}$, and $\alpha_{ML2}$ with respect to pL (= -log [L]) can be shown, by methods developed in Chapter 8, to be

Using an ethylenediamine -$Cu^{2+}$ titration as a model, we can adapt Eq. 9-3 to obtain the titration curve equation and its derivative as follows:

$$F = (\alpha_{CuL} + 2\alpha_{CuL_2}) + \frac{V_{Cu} + V_L}{C_{Cu}V_{Cu}}[L]$$

$$\frac{dF}{dpL} = -2.303 \left\langle (\alpha_{Cu}\alpha_{CuL} + \alpha_{CuL}\alpha_{CuL_2} + 4\alpha_{Cu}\alpha_{CuL_2}) + \frac{V_{Cu} + V_L}{C_{Cu}V_{Cu}}[L] \right\rangle$$

This equation properly bears a striking resemblance to the analogous Equation 8-17 and can be very useful in describing the metal buffer index as well as for the sharpness index in complexometric titrations.

## PROBLEMS

**9-1** Fifty milliliters of a 0.001 M $Cu^{2+}$ solution, buffered by 0.02 M $NH_3$ and 0.04 M $NH_4Cl$ is titrated with 0.005 M $Na_2H_2Y$(EDTA). Construct the titration curve, the difference titration curve, and by using the expanded $V_Y$ scale (9.9 to 10.1), read values of pCu when the error in $V_Y$ is $\pm 0.1$, 0.3, and 0.5%.

Repeat the calculations for conditions described above, but with no $NH_3$ present (pH the same as before). Plot both curves in the same graph (see p.8.25).

**9-2** Fifty milliliters of 0.002 M $Zn^{2+}$ solution, buffered at pH 9.0 with a noncomplexing buffer (e.g., borate) is titrated with 0.005 M Trien. Construct the titration curve, the difference titration curve, and by using the expanded $V_L$ scale (19.9 to 20.1), read values of pZn when the error in $V_Y$ is $\pm 0.1$, 0.3, and 0.5%.

**9-3** Calculate the conditional $\beta_{MIn}$ values for each of the metallochromic indicators in Table 9.1 for $Cu_{2+}$ and $Zn^{2+}$ at the pH values in the two problems above. At pM values equal to the corresponding log $\beta_{MIn}$ values, what are respective titration errors?

# Chapter 10

# Titrations III
# Redox Titrations

Titrations based on oxidation-reduction reactions enjoy wide use. Permanganate, dichromate, and iodine and iron(II), tin(II), thiosulfate, and oxalate are commonly used oxidizing and reducing titrants, have been employed to determine components in both inorganic and organic analysis. As we saw in Chapter 7, solvent water does not play as central a role as in acid-base titrations. Oxidants or reductants strong enough to decompose water are not practical as titrants.

### Derivation of Redox Titration Curve Equations

Just as acid-base reactions are governed by proton exchange, redox reactions are characterized by electron exchange. The balance equation analogous to the PBE, however, is best described by equating the amount of reactant per single electron transferred. This allows us to take into account the different numbers of electrons gained or lost by various oxidants and reductants. The equation for the general case of redox titration can be written as

$$n_B Red_A + n_A Ox_B \rightleftharpoons n_B Ox_A + n_A Red_B$$

where system A involves transfer of $n_A$ electrons and system B, transfer of $n_B$ electrons. The composition of a solution initially containing $V_R$ mL of $C_R$ M of the reduced form of a redox couple A to which has been added $V_o$ mL of $C_o$ M of a titrant which is the oxidized form of a redox couple B, reacts in accord with an "electron balance equation," EBE, given by

$$V_A[Ox]_A/n_A = V_B[Red]_B/n_B \qquad (10-1)$$

Thus, at any stage during the titration the reduced form of the titrant and the oxidized form of the system being titrated are **equivalent**, i.e.,

$$n_B[Ox]_A = n_A[Red]_B \tag{10-1a}$$

Mass balance for A and B:

$$[Red]_A + [Ox]_A = C_R V_R/(V_R+V_o)$$

and

$$[Red]_B + [Ox]_B = C_o V_o/(V_R+V_o)$$

Utilizing the treatment developed in Chapter 7 to describe concentrations in terms of **f** fractions, we have

$$f_{O(A)} = \frac{\pi^{n_A}}{(\pi^{n_A} + K_A^{n_A})} \quad f_{R(A)} = \frac{K_A^{n_A}}{(\pi^{n_A} + K_A^{n_A})}$$

$$\text{and} \quad f_{O(B)} = \frac{\pi^{n_B}}{(\pi^{n_B} + K_B^{n_B})} \quad f_{R(B)} = \frac{K_B^{n_B}}{(\pi^{n_B} + K_B^{n_B})} \tag{10-2}$$

and substituting in 10-1a

$$9n_B \frac{C_o V_o}{V_o + V_R} \cdot \frac{\pi^{n_A}}{\pi^{n_A} + K^{n_A}} = n_A \frac{C_R V_R}{V_o + V_R} \cdot \frac{K^{n_B}}{\pi^{n_B} + K^{n_B}} \tag{10-3}$$

## Calculations of pE for Selected Individual Points

Selected points along the titration curve are calculated in a manner that closely resembles those we used in the last two chapters. With oxidation-reduction titrations, the method is even simpler when we recognize that on either side of the equivalence point there is an excess of one of the two redox couples, allowing us to calculate the concentration ratio of oxidized to reduced forms of the substance being titrated or of the titrant. Knowing this ratio, which we get from stoichiometric considerations, permits us to solve the Nernst equation for the pE value.

**Example 10.1**

What are the pE values of a 50 mL portion of a solution of 0.005 M $SnCl_2$ when (a) 5 mL (b) 60 mL of a 0.01 M $FeCl_3$ titrant have been added? The two points have been selected so that they typify calculations on each side of the equivalence point.

(a)    The total Sn concentration is given by $50 \times 0.005 = 0.25$ mmol. When 5 mL of titrant has been added, 10% of $Sn^{2+}$ has been oxidized to $Sn^{4+}$, and virtually all the $Fe^{3+}$ has been reduced to $Fe^{2+}$. Thus, while one solution can have only one pE value, meaning that pE calculated either from the Sn or the Fe Nernst equations must be the same, we have information for solving only the first (What can we say about $[Fe^{3+}]$ being almost zero?)

$$pE_{Sn} = pE° + \tfrac{1}{2}\log([Sn^{4+}]/[Sn^{2+}])$$
$$= 2.6 + \tfrac{1}{2}\log(\tfrac{1}{2}\times5\times0.01)/(0.25 - \tfrac{1}{2}\times5\times0.01) = 2.1$$

Note that in the logarithm, the amount in mmoles rather than the concentrations were used. The solution volumes cancel out in a ratio. Hence, except for very dilute solutions and, of course, activity coefficient changes, pE will be independent of total concentrations.

(b)    When 60 mL titrant has been added, $Fe^{3+}$ is in excess, and the $[Sn^{2+}]$ is virtually zero, making use of the Fe Nernst equation necessary.

$$pE = pE° + \log([Fe^{3+}]/[Fe^{2+}])$$
$$= 12.0 + \log(60\times0.01 - 2\times50\times0.005)/(2\times50\times0.005) = 11.31$$

Once the pE value is found for each point, one can use this value to learn what "virtually zero" $[Fe^{3+}]$ in (a) or $[Sn^{2+}]$ in (b) really is. For the latter

$$11.3 = 2.6 + \tfrac{1}{2}\log(0.25/(60 + 50)/[Sn^{2+}]$$
$$\text{or } \log[Sn^{2+}] = 2(2.6 - 11.3) + \log(0.25/(60 + 50) = -15.35$$
$$\text{that's low!}$$

The equivalence point pE value can be obtained another way. At the equivalence point: $n_B V_R C_R = n_A V_0 C_0$ , which means

$$\frac{\pi^n}{K_B^{n'}} \cdot \frac{(\pi^{n'} + K_B^{n'})}{(\pi^n + K_A^n)} - 1$$

*This leads to*                                                                                (10-4)

$$\pi^{(n_A + n_B)} + \pi^{n_A} K_B^{n_B} - \pi^{n_A} K_B^{n_B} + K_A^{n_A} K_B^{n_B}$$

*or:*        $\pi^{(n_A + n_B)} - K_A^{n_A} \cdot K_B^{n_B}$

Expressed in logarithmic form

$$(n_A + n_B) pE_{equiv.pt.} = n_A pE_A^\circ + n_B pE_B^\circ$$

or

$$pE_{equiv.pt.} = (n_A pE_A^\circ + n_B pE_B^\circ)/(n_A + n_B) \tag{10-5}$$

Thus, the pE at the equivalence point is the **weighted average** of the pE° values of the two redox couples.

## Redox Indicators

A redox indicator is a redox couple whose forms have distinct and characteristic spectra of high molar absorbances. They are frequently acid/base couples as well, making pH control necessary. They are useful in conducting titrations where only the stoichiometry and not the potential values are needed. Several are described below.

### Table 10.1  Selected Oxidation Reduction Indicators

| NAME | pE⁰ IN 1 M $H_2SO_4$ | Oxidized/Reduced Colors |
|---|---|---|
| Iron(II) Chelate of | | |
| 1,10-Phenanthroline(phen) "Ferroin" | 1.11 | Pale blue/Red |
| 5-Nitro-phen | 1.25 | Pale blue/Red |
| Diphenylamine Sulfonic Acid | 0.85 | Red-violet/Colorless |
| Methylene Blue | 0.53 | Blue/Colorless |
| Tris(2,2'-bipyridyl)Fe | 1.12 | Pale blue/Red violet |
| Tris(2,2'-bipyridyl)Ru | 1.29 | Pale blue/Yellow |

**Example 10.2**

Develop the titration curve for 50 mL of 0.005 M $SnCl_2$ with $V_0$ mL of 0.01 M $FeCl_3$. In terms of Equation 10-2, $n_A = 2$, $n_B = 1$, $C_R = 0.005$, $V_R = 50$, $C_0 = 0.01$, $\pi = 0^{pE}$, $K_A = 10^{2.6}$, and $K_B = 10^{12.0}$.

In column **A** enter value of E from 0 to 17 in steps of 0.1. In column **B**, enter $\pi$ using the formula + **10^A2**. In column **C** enter the formula for $V_0$ in Equation 10-2, using absolute addresses for $n_A$, $n_B$, $C_R$, $V_R$, $C_0$, $K_A$, and $K_B$. (In this way, this spreadsheet problem can be used for any redox titration curve.) The graph will use **C2..C171** as Series 1 and **B2..B171** as the X series. (Figure 10.1)

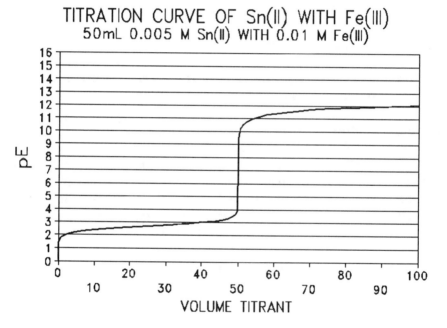

**Figure 10.1 Titration Curve based on Equation 10-2**

**Example 10.3**

Using the graph prepared in Example 10.2, locate both the volume and pE at the equivalence point by using a difference curve, i.e., in column **D**, enter the formula +**(B3 - B2)/(C3 - C2)**. Compare pE at the equivalence point with value obtained in Equation 10-3.

# PROBLEMS

**10-1**   Develop the curve for the titration of 10 mL of 0.10 M $FeSO_4$ in a 1.0 M $H_2SO_4$ medium with 0.05 M $Ce(SO_4)_2$. Obtain the difference curve. Compare the equivalence point potential obtained from the curve to that calculated from Equation 10-3.

**10-2**   Develop the curve for the titration of 25 mL of 0.09343 M $Na_2S_2O_3$ with 0.1216 M $I_2$. Compare the equivalence point potential with the potential of the mixture when an excess of 1 drop (0.03 mL) of the $I_2$ solution has been added.

**10-3**   Calculate the error in mL that would be incurred if the $pE°$ of the best redox indicator in Table 10.1 for the titration in 10-1 and then that in 10-2 were taken as the equivalence point $pE$ value?

**10.4**   Develop the titration curve for the titration of 50 mL of 0.01 M ascorbic acid with (a) 0.02 M $K_3Fe(CN)_6$ or (b)0.02 M $FeCl_3$ using the spreadsheet to obtain $pE$ values at increments of 0.025 $pE$ units. Compare the slopes of these two curves in the vicinity of the equivalence point, i.e., $V_{equiv. pt.} \pm 2.0$ mL.

# Chapter 11

# Titrations IV
# Precipitation Titrations

---

Although titrations involving precipitation reactions are not as widely used since the introduction of complexation titration continue to be useful for determination of anions, particularly the halides.

As with the other reaction types, the precipitation titration curve can be described by a single equation. If $V_X C_X$ of a chloride solution is titrated with $C_{Ag}M$ $AgNO_3$, then at any point after the addition of $V_{Ag}$ mL, balance equations can be written for $[Cl^-]$ and $[Ag^+]$. The total Ag content is distributed between dissolved $Ag^+$ and precipitated AgCl. Similarly, total Cl is the sum of dissolved $Cl^-$ and precipitate.

$$C_{Ag}V_{Ag} = [Ag^+](V_{Ag} + V_X) + \text{mmol AgCl}$$

$$C_X V_X = [Cl^-](V_{Ag} + V_X) + \text{mmol AgCl}$$

which gives, after eliminating mmol AgCl between the two equations,

$$C_{Ag}V_{Ag} = (V_{Ag} + V_X)([Ag^+] - [Cl^-]) + C_X V_X$$

then

$$V_{Ag}(C_{Ag} - [Ag^+] + [Cl^-]) = V_X(C_X + [Ag^+] - [Cl^-])$$

or

$$or \quad V_{Ag} = V_X \frac{(C_X + [Ag^+] - [Cl^-])}{(C_{Ag} - [Ag^+] + [Cl^-])} \qquad (11-1)$$

*Generally, for a titration yielding the precipitate, $M_x X_m$*

$$V_M = V_X \frac{(xC_x + [M^{m+}] - [X_{x-}])}{(mC_M - [M^{m+}] + [X^{x-}])} \qquad (11-2)$$

196

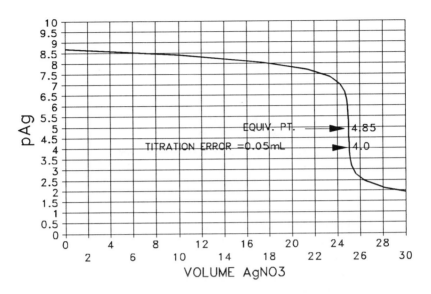

**Figure 11.1 Titration of 50 mL 0.1 M NaCl & 0.2 M AgNO₃**

The Cl- concentration can eliminated from Equation 11-1 since, in the presence of AgCl, $[Cl^-] = Ksp/[Ag^+]$. The formal resemblance of this equation with Equation 8-2 for strong acid-base titrations is interesting. Note that both AgCl and HOH are similar in that the ion product in each case is a constant.

Equation 11-2 describes the most general case of precipitate titration, one in which the stoichiometric relation of anion and cation are not limited to 1:1.

A spreadsheet for the Ag - Cl titration can be prepared by filling column A with values of pAg from 0 to 8 in steps of 0.25, and successive columns with $[Ag^+]$, $[Cl^-]$, and $V_{Ag}$ using the appropriate formulas just developed. Figure 11.1 was prepared for the case in which 50 mL of 0.1 M NaCl is titrated with 0.2M AgNO₃. Use $K_{sp}(AgCl) = 10^{-9.70}$

The pAg equivalence point in the Ag⁺ - Cl⁻ titration, like the pH of the strong acid-strong base, is independent of the initial concentrations of the reactants.

$$pAg_{equiv} = \tfrac{1}{2} pK_{sp} = 4.85$$

## Mohr Titration

In the Mohr titration, the appearance of a red precipitate, $Ag_2CrO_4$, is used to indicate the endpoint. If $K_{sp}(Ag_2CrO_4) = 10^{-10.5}$, what is the $[CrO_4^{2-}]$ that will eliminate endpoint error?

$$[Ag^+]^2[CrO_4^{2-}] = 10^{-10.5}$$

$$[CrO_4^{2-}] = 10^{-10.5}/10^{-2(4.85)} = 10^{-0.8}$$

$$= 10^{-0.8} = 0.16 \ M$$

This is a rather concentrated chromate solution, making this impractical to use. By examining the region of greatest pAg change in the titration curve, we should be able to select a point that can be reached with a lower $[CrO_4^{2-}]$ without incurring an unacceptable titration error. This is conveniently done with the spreadsheet-generated graph by expanding the X scale so that it extends from -49.5 to 50.5, the pAg at 50.05 mL drops to 4.0, and, at 50.10 mL, pAg = 3.75. If we find a +0.05-mL titration error acceptable (from the graph, pAg = 4.0) the chromate concentration required as indicator in this case is

$$[CrO_4^{2-}] = 10^{-10.5}/10^{-2(4.0)} = 10^{-2.5} = 0.0033 \ M$$

This can be achieved by adding 1.7 mL of 0.1 M $Na_2CrO_4$ to the 50 mL of NaCl at the outset of the titration.

$$0.1 \times V_{In} = 0.0033 \times 50$$

If, as good analytical practice demands, the standardization titration also contains the same amount of indicator solution as each of the determination runs, then the indicator errors will be almost negligible.

The strategy just employed for obtaining titration errors directly from the expanded scale titration curve represents another of the many advantages and freedoms resulting from the use of spreadsheets in calculations.

### Titration of Mixtures of Halides

When a solution containing a mixture of halide ions is titrated with $Ag^+$, the first precipitate will be, unless there is really great disparity in the concentrations of the halides, AgI. When the $I^-$ is substantially totally precipitated, the $Br^-$ and $I^-$ will follow. Figure 11.2 represents a typical

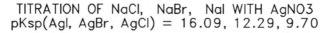

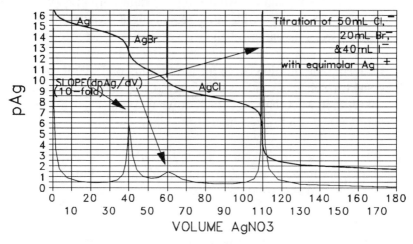

**Figure 11.2 Titration of NaCl, NaBr, NaI with AgNO$_3$**

titration curve of such a mixture.  It was derived from an equation that is
an adaptation of 11-1.  Equation 11-3 relates the total volume of the Ag$^+$,

$$V_{Ag} = \frac{C_{Cl}V_{Cl}}{C_{Ag} - [Ag^+] + [Cl^-]} + \frac{C_{Br}V_{Br}}{C_{Ag} - [Ag^+] + [Br^-]} + \frac{C_I V_I}{C_{Ag} - [Ag^+] + [I^-]}$$

(11-3)

titrant to the sum of volumes required for each of the halides present.  It
should be noted that the concentrations of halides in the equation are those
of the **added** solutions and not in the final, mixed solution.  In other words,
$C_X V_X$ represents the amount in millimoles of the added halide.

## PROBLEMS

**11-1**  With the help of Equation 11.2, develop the titration and difference
curves for the titration of 35mL 0.05 M Na$_2$SO$_4$ with 0.05 M
Pb(NO$_3$)$_2$.

**11-2**  Using Equation 11-3, develop a series of simulated precipitation titration curves in which $Ag^+$ is used to titrate three singly charged anions, $X^-$, $Y^-$, and $Z^-$, whose log $K_{sp}$ values are -12, -9, and -6. Repeat the exercise with other sets of log $K_{sp}$ values. How does the log $K_{sp}$ interval affect the ease of endpoint detection?

**11-3**  Develop the titration curve for the titration of 25.0 mL of 0.10 M NaCl which also contains 0.025 M $K_2CrO_4$ using the spreadsheet. With 0.10 M $AgNO_3$ as titrant, at what volume($\pm$ 0.01 mL) will $Ag_2CrO_4$ begin to precipitate?

**11-4**  Develop the titration curve for the titration of 50 mL of 0.05 M $CaCl_2$ buffered at pH 10.0 with an identically buffered 0.05 M phosphate titrant solution.

**11-5**  For the titration curves developed in the previous examples, calculate the equivalence point for each of the intermediate and final points. Further, add an additional set of pM values at 0.01 increments within the range of $pM_{equiv.\ pt.}$ $\pm$ 0.3(remember to sort before regraphing) and find, by inspecting the resulting diagram visually, what the pM values are when the titration volume errors are 0.1, 0.5, and 1%.

# Chapter 12

# Statistical Treatment of Data

Most of the topics covered in this text are of interest to many others as well as to analytical chemists. Of these, however, statistical concepts are clearly the most generally applicable and widely encountered. For all experimental scientists, the evaluation of reliability of their results is of paramount interest. Except for pure numbers, numerical values are virtually meaningless when not accompanied by a specified estimate of their reliability.

In addition to evaluation of new methods (how well do the analyses agree with each other as well as with older methods), or new people's ability to follow established methods (how well do student data agree with it), questions about the validity (i.e., representativeness) of samples must be addressed not only by analytical chemists, but all experimental scientists and engineers, by epidemiologists, marketers, journalists, politicians, and others interested in opinion surveys.

Let us begin by defining some essential terms dealing with expressions of describing central tendency and scatter in data, of reliability, precision, and accuracy.

### Mean, Variance, and Standard Deviation

When a measurement is replicated, the set of n values is expressed as a **mean**, or **average**, x, i.e., the sum of values divided by their number,

$$mean = \bar{x} = 1/n \sum_i^n x_i \qquad (12\text{-}1)$$

The spread or dispersion of this set of values is measured by the **variance**, V, which is the sum of the squares of the deviation of each value from the mean, divided by one less than the number, n, of values in the set.

$$V = \sum_i^n (x_i - \bar{x})^2/(n-1) \qquad (12\text{-}2)$$

The positive square root of the variance is called the **standard deviation**, and is widely used.

In some disciplines, the variance is defined by

$$V' = \sum_i^n (x_i - x)^2/n$$

$$s - \sqrt{V} - \sqrt{\frac{\sum_i^n (x_i - \bar{x})^2}{n-1}} \tag{12-3}$$

from which the corresponding standard deviation, $s'$, is defined.

$$s' - \sqrt{V'} \tag{12-4}$$

Generally speaking, analytical chemists employ V and s. Unfortunately, however, spreadsheet programs such as Quattro and Lotus 123 employ $V'$ and $s'$ necessitating a correction, e.g., for s

$$s - s'\sqrt{n/(n-1)} \tag{12-5}$$

Notice that s is larger than $s'$. At what value of n does this difference become 10%, 5%, 1%? (Show that the n values are 6, 11, and 51, respectively.) Obviously, as n increases, the differences in s and $s'$ (as do V and $V'$) become negligible.


## Precision and Accuracy

The first estimate of the reliability of the mean value of a set of analytical data of a specific sample is obtained by seeing how closely the individual values agree. This is expressed in terms of the standard deviation. There are several aspects of precision that are worth noting that arise in comparing performances among operators under the same and varied conditions. These are called repeatability and reproducibility.

Accuracy is a term describing deviation of an observed value (or the mean of observed value) from the "true" or, more realistically, the generally acceptable value. For example, the "true value" of a parameter (e.g., %Ni, %Fe, %S) of a well-defined sample is the mean value obtained by the work of several teams of experienced, competent analytical chemists. The National Institute of Standards and Technology, formerly National Bureau of Standards, make available to the analytical chemistry community a series of carefully prepared materials analyzed by several kinds of established methods, known as Standard Reference Materials, including metals, minerals, and environmental samples.

"True" values of fundamental quantities, such as atomic masses, etc., which must be experimentally measured, can best be evaluated in terms of an analysis of all of the sources of error inherent in the experimental method. Such analysis requires detailed knowledge of the chemistry of the systems involved (e.g., equilibrium, kinetic factors) to gauge the effect of solubility, incomplete or side reactions, etc. on the result. These topics form an important aspect of analytical chemistry.

Neither of these terms attempts to describe variations arising from sample to sample. Indeed, with many problems facing contemporary analytical chemists, difficulties in obtaining a representative sample of heterogeneous materials can cause errors far larger than those arising from uncertainties in executing the analytical methods. Methods of obtaining proper samples are complex, however, and will be discussed later.

**Repeatability** is the closeness of agreement between successive results obtained with the same method on identical test material, under the same conditions (same operator, same apparatus, same laboratory, and short intervals of time). The value below which the absolute difference between two single test results obtained in the above conditions may be expected to lie with a specified probability.

**Reproducibility** is the closeness of agreement between individual results obtained with the same method on identical test material but under different conditions (different operators, different apparatus, different laboratories, and/or different times). The value below which the absolute difference between two single test results on identical material obtained by operators in different laboratories, using the standardized test method, may be expected to lie with specified probability.

## Classification of Errors

The results of any experiment are subject to three types of errors: determinate, indeterminate, and accidental (or gross). **Determinate errors** are those for which we can attribute a reason, notably a specific shortcoming in the method or apparatus. For example, when the precipitation washing and weighing of AgCl is employed to estimate the chloride content of a dissolved sample, the amount of chloride lost to the very small but predictable solubility of AgCl as well as the contamination of the precipitate (a phenomenon known as coprecipitation) results in errors. These are predictable and, in principle, correctable when, as usual, the density of the object being weighed is lower than that of the weights. The observed weight will be in error unless the buoyancy correction is applied.  In titrimetric analysis, determinate errors arise if the indicator used signals completion of the titration reaction before or after being the true equivalence point. An important reason for learning the underlying principles governing reactions involved in analysis (this amounts to all of chemical science) is to enable us to assess and correct for all determinate errors.

At the other extreme are the totally arbitrary and unexpected category of **accidental** or **gross errors**.  Mistakes in recording values, use of dirty equipment, unrecognized losses or contamination of samples represent unfortunate sources of error that we can only hope are minimized by the training, discipline, and experience of the analyst.  Although such errors, by their very nature, can neither be anticipated or corrected, we can hope to recognize such occurrences by statistical analyses.  Of course, multiplicate analyses should serve to reveal such errors.

The remainder of this chapter will be devoted to **indeterminate errors**, those for which we cannot assign a specific reason but which represent the very limits of our ability to observe and the limits of the instruments we employ.  If an analytical balance can detect a change of 0.0001 g, each measurement will be uncertain at least to this extent.  We term the uncertainty **random** because it is as likely to be negative as positive. The magnitude of the error is much more likely to be small; the probability of occurrence of an error falls with the size of the uncertainty.

## The Gaussian Error Function

The centerpiece of statistical treatment of data is the Gaussian Error function which gives mathematical expression for the behavior of random errors, namely, that (a) errors are just as often positive as negative and (b) smaller errors are more probable than larger ones

$$y = \frac{1}{\sigma\sqrt{2\pi}} e^{-(x-\mu)^2/2\sigma^2} = \frac{1}{\sigma\sqrt{2\pi}} e^{-z^2/2} \qquad (12\text{-}6)$$

where $\mu$ represents the mean value of x, and $\sigma$, the standard deviation, for an **infinite** number of samples; z represents each deviation in units of $\sigma$, permitting the use of a single graph to describe the Gaussian curve.

### Example 12.1  Properties of the Gaussian Error Curve

In column A, labeled as SIGMA, block fill A2..A1003 values from -5 to +5 in steps of 0.01.  In B, enter the values of Y2 by means of the formula in Equation 12-6 using a value of $\sigma = 1$; fill the rest of B using block copy.  Generate an XY graph using B2-B1003 as Series 1, A2-A1003 as the X series.  Note the familiar bell-shaped curve.

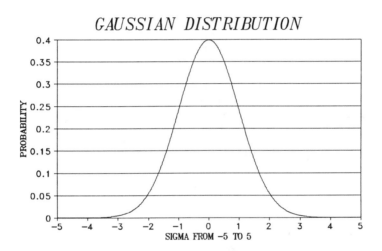

## GAUSSIAN DISTRIBUTION

**Figure 12.1**  Graph of the Gaussian Function in 0.1 Increments of Sigma

Let us compare the shape of this theoretical curve with that obtained by taking a frequency distribution of 1000 random numbers.  In column C, label C1 as RAND, then in C2 enter the formula +0.4(1+@RAND-@RAND) using /**BC** to copy the result in C3..C1003.  In order to obtain a frequency distribution of these values, let us use column D to contain the contents of column C <u>with values only</u>.  To do this, use the instruction, /**EV**{/**BAV**; /**RV**} (you will not see a difference in the corresponding cell contents of C and D, but you will notice that each cell in C has a corresponding value of the initial formula while the D cell will just list the pure number).  Before initiating the frequency distribution, you must first set up a series of intervals referred to as "bins".  For this purpose, use column

E and after labeling E1 as BINS, use **/EF{/BF; /F}** to fill block F2-F41 with a series of numbers starting from 0, proceeding in increments of 0.025. Using the command, **/TF{/AF; /DD}**, you will be prompted for the block of values that you want to analyze. The block to be examined is D2..D1003 and the "bin" location is F2..F41. (Press enter.) The distribution results will be displayed in F2..F41. Display this as a bar graph, with E2..E41 as the X-series and F2..F41 as the first series.

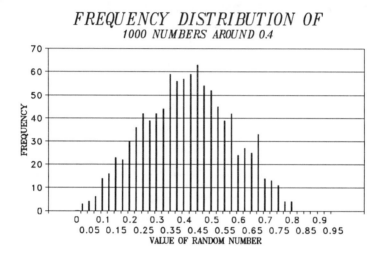

**Figure 12.2**  Frequency Distribution of 1000 Random Numbers

Note that this graph approximates the Gaussian curve, which is a continuous function, so that it is comprised of an infinite number of points. Repeat the frequency distribution using 0.05 as bin interval. With the smaller number of bins, the distribution will conform better to the shape of the error curve. Why?

## Confidence Intervals and Confidence Limits

Using the Gaussian curve, the probability of obtaining a result within a certain interval around the mean can be related to the area under the curve within that interval. First let us obtain the area under the entire curve from

-5$\sigma$ to +5$\sigma$. Remember that the area of a rectangle is the product of height times width. In the spreadsheet we have divided the entire curve into 1000 rectangles each having a width of 0.01 and a height of y, given by the equation for the Gaussian curve and listed in column B. Since the width of each of these rectangles is constant (0.01) the area for each is proportional to the y value. To get the entire area, simply use the function @SUM(B2..B1003), locating the answer in a convenient cell, e.g., F2. To find the area of the curve between ±1$\sigma$, 2$\sigma$, and 3$\sigma$ simply locate the appropriate intervals in column A, and find the areas by @SUM(B401..B602), (B301-B702), and (B201-B802), respectively, using cells F3, F4, and F5. In cells G3 to G5 place the results of dividing each of these areas by the total area in F2 multiplied by 100. The percent area under the error curve is exactly equivalent to the percent probability of finding a single additional value between ±1$\sigma$, 2$\sigma$, and 3$\sigma$ ($\sim$68, 96, and 99.7%, respectively).

The intervals for values included around the mean (±1$\sigma$, 2$\sigma$, etc.) are called **confidence intervals** and the percent probability of finding values within such a limit is termed **confidence level**. A useful graphical display of the relationship between these two quantities may be seen in Figure 12.3.

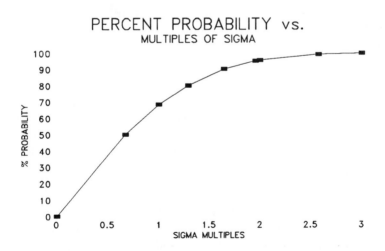

**Figure 12.3 Confidence Levels at Various Confidence Limits**

For example, in a data set describing a copper alloy having a mean ($\mu$) of 41.26 $\pm$ 0.12 % (standard deviation) Cu, any future samples of the same alloy will yield results that will fall between 41.02 and 41.50% Cu (confidence interval is $\pm 2\sigma$) with a 96% probability (confidence level). An important corollary is that the chances that values **outside** these limits are of indeterminate origin is very small (100 - 96%) or 4%. Such values may either be discarded as invalid, even if the cause is unknown. Alternatively, such values can indicate that the samples were different from the rest with a 96% probability. We shall be using these concepts throughout the chapter. With this introduction, we can now deal with questions concerning reliability.

## Statistical Analysis of Small Sets of Data

The Gaussian error function correctly describes the variations to be expected in a population of values of infinite size, an obviously unattainable level of replication in practice. With the general guidelines embodied in this function, a body of statistical rules applicable to small sets of data has been developed.

This basic statistical relationship has been modified to accommodate the more practical situation when small sets of data are involved. In the analysis of a sample, it is not only more expensive to increase the number of replicate determinations, but this consumes more of the sample. The expression involves the use of t, a factor by which the standard deviation, s, is multiplied to define the confidence limit in which the true value, $\mu$, will be found.

$$\mu = \bar{x} \pm ts/\sqrt{N-1} \qquad (12\text{-}7)$$

where N is the number of values used to find the mean, x, as seen in Table 12.1, the value of t decreases rapidly with N, as expected from the increase of reliability of measurements with the number of replicates.

**Table 12.1  Values of $t_{CL}$ at Various Desired Confidence Levels (CL) As a Function of N, the Number of Determinations in a Set**

| N | $t_{90}$ | $t_{95}$ | $t_{99}$ |
|---|---|---|---|
| 2 | 6.31 | 12.71 | 63.7 |
| 3 | 2.92 | 4.30 | 9.92 |
| 4 | 2.35 | 3.18 | 5.84 |
| 5 | 2.13 | 2.78 | 4.60 |
| 6 | 2.02 | 2.57 | 4.03 |
| 7 | 1.94 | 2.45 | 3.71 |
| 8 | 1.90 | 2.36 | 3.50 |
| 9 | 1.86 | 2.31 | 3.36 |
| 10 | 1.83 | 2.26 | 3.25 |
| 12 | 1.80 | 2.20 | 3.11 |
| 14 | 1.77 | 2.16 | 3.01 |
| 20 | 1.73 | 2.09 | 2.86 |
| 30 | 1.70 | 2.04 | 2.76 |
| 40 | 1.68 | 2.02 | 2.70 |
| 60 | 1.67 | 2.00 | 2.66 |
| $\infty$ | 1.64 | 1.96 | 2.58 |

Notice also how different $t_{90}$ is from $t_{99}$ when N is small (10:1 at N = 2) but much less so as N increases (3:1 at N = 3 to 1.6:1 at N = 60).

Table 12.1 also provides a guide for rejecting an outlier among a set of data. For example, if one of a set of four sample results seems to be out of line with the others, simply obtain the mean and standard deviation for the 3 remaining results. At a 95% confidence level the fourth result may be safely discarded if its deviation from the mean is 4.30 s or greater.

## Comparison of Two Means

There is frequent need to compare two or more sets of data. For example, sample uniformity can be estimated by analysis of sets of laboratory samples gathered from different parts or at different times or both of the object(s) being measured. Although the means of the various sets can be expected to differ somewhat, how much of a difference indicates nonuniformity or _real_ difference rather than experimental uncertainty.

Let $X_a$ represent the mean of a set of $N_a$ determinations and $X_b$ of another set of $N_b$. To test the significance of the difference, $| \ X_a - X_b \ |$ , first calculate the pooled standard deviation of the group, $s_p$, where

$$s_p = \frac{\sqrt{(N_a-1)s_a^2 + (N_b-1)s_b^2}}{N_a + N_b - 2} \qquad (12\text{-}8)$$

The criterion for judging that $X_a$ and $X_b$ are significantly different if

$$| \ X_a\text{-}X_b \ | \ > \ ts_p \ \sqrt{(N_a+N_b)/N_aN_b}$$

where values of t are obtained from Table 12.2.

Table 12.2 Expected Difference Between Two Means, $X_a$ and $X_b$, for Various Confidence Levels

Confidence Level, or Probability That $X_a - X_b$ is Less Than the Limits Above When $\mu_a = \mu_b$

| $N_a + N_b$ | $t_{90}$ | $t_{95}$ | $t_{99}$ |
|:---:|:---:|:---:|:---:|
| 3 | 6.31 | 12.71 | 63.7 |
| 4 | 2.92 | 4.30 | 9.92 |
| 5 | 2.35 | 3.18 | 5.84 |
| 6 | 2.13 | 2.78 | 4.60 |
| 7 | 2.01 | 2.57 | 4.03 |

| 8 | 1.94 | 2.45 | 3.71 |
|---|------|------|------|
| 9 | 1.90 | 2.36 | 3.50 |
| 10 | 1.86 | 2.31 | 3.36 |
| 20 | 1.73 | 2.10 | 2.88 |
| 30 | 1.70 | 2.05 | 2.76 |
| 40 | 1.69 | 2.02 | 2.71 |
| 60 | 1.67 | 2.00 | 2.66 |
| ∞ | 1.64 | 1.96 | 2.58 |

## Propagation of errors

A common scientific activity is the combination of experimental values to calculate a desired parameter. To take a simple example, in order to standardize a ~0.1-M HCl solution, we would use an aliquot of a solution prepared by dissolving a given weight of primary standard $Na_2CO_3$ in a known volume of water. Titration of this HCl solution with the $Na_2CO_3$ would yield a volume of acid, which together with the weight of $Na_2CO_3$, is used for the standardization calculation. How can the effect of various errors arising in the different measurements (here, mass, and volume) on the reliability of the calculated value (here, M) be estimated. We will not consider the effect of either determinate or accidental errors. With indeterminate or random errors, the generally accepted rule is that the estimated variance of the overall calculation is the sum of the variances of the individual steps.

$$s^2_{overall} = \sum_i s_i^2 \qquad (12\text{-}9)$$

The standard deviation, therefore, is the square root of the sum of the squares of the component standard deviations.

$$s_{overall} = \sqrt{\sum_i s_i^2} \qquad (12\text{-}10)$$

## The Noise Function

Sometimes it is useful to be able to alter  the results of a simulated experiment by building random errors into the data set at a controlled level.

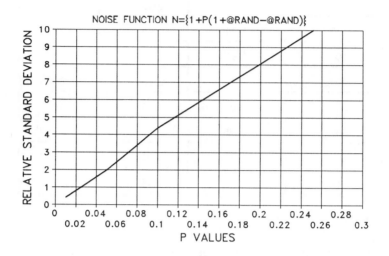

**Figure 12.4** Relation of Noise Function to Relative Standard Deviation

This can be easily and conveniently accomplished with the QPRO random function.

Let us define a noise function.

$$N = \{1 + P(1 + \textbf{@RAND} - \textbf{@RAND})\} \qquad (12\text{-}11)$$

This function can be used to introduce a desired level of relative uncertainty (Figure 12.4) to any set of data by using it as a multiplier.

## PROBLEMS

**12-1**   From a set of analyses for which we have calculated a mean, x, and a standard deviation, s, what can we say about the chances of the results of further experiments on the same sample giving the same value as x; what is the probability of the value falling between x + ts and x - ts?

12-2    A series of samples taken from a single piece of Cu ore were found to have 0.412, 0.422, 0.409, 0.430, 0.417, 0.413, 0.433, 0.431, 0.422, and 0.429 %Cu. Should all 10 of these results be used in reporting the sample composition? What is the range of values of %Cu for this sample to the 95% confidence level? If it is known that the analytical method used here has a precision of 0.6 relative %, are the results indicative of sample homogeneity?

12-3    Equation 12-11 was developed to deal with random relative error (noise). Develop an appropriate equation suitable to generate random absolute errors at various levels. Use both to examine the absolute standard deviation of a measurement of mass of 0.0002 g of a substance whose purity is known to 1%, assuming a weighing error of $\pm$ 0.0002 g. Obtain 20 values using randomization of each factor and 20 values with both factors employed. What are the standard errors of the resultant means?

# Chapter 13

# Spectrometry

---

## Interactions of Matter
## and Electromagnetic Energy

Probably one of the most fascinating phenomena in science is the interactions of light, electromagnetic radiation, with matter. The essential nature of matter, the nature of the forces acting between the parts of atoms, between the atomic components of molecules, is revealed by such phenomena. It produces a cosmic arithmetic that relates the energy changes associated with changes in atomic and molecular structural states. These phenomena provide the basis for one of the most widely used methods of chemical analysis: colorimetric and spectrophotometric analysis.

Indeed, the principal way that energy changes associated with such structural changes are understood and measured is by means of the characteristic electromagnetic light energy that is either produced (emitted) or consumed (absorbed). The electromagnetic energy is given by

$$\Delta E = h\nu$$

where h is called Planck's constant, $6.624 \cdot 10^{-27}$ erg·sec, and $\nu$ is the frequency which is also describable as $c/\lambda$ (C is the speed of light, $3 \times 10^{10}$ cm/sec., and $\lambda$ is the wavelength of the light).

Electromagnetic energy ranges from very high frequencies of $\gamma$-rays associated with nuclear reactions, through X-rays associated with transitions of the inner shell electrons, the UV and visible range involving usually outer-shell electron transitions, to the near and far infrared regions resulting from molecular vibrations and rotations (r.f. microwave?).

The mighty and mysterious quantum theory explains the interaction of light and matter in terms of well-defined energy states that characterize an atom or molecule. When the atom or molecule absorbs energy to enter a higher energy state, the energy used is

$$\Delta E = h\nu = hc/\lambda \qquad (13\text{-}1)$$

where $v$ is the frequency of the energy absorbed, and the corresponding wavelength. When the material changes is from higher to lower energy, energy is (usually) emitted as light, i.e., electromagnetic energy as described by Equation 13-1.

## Origin of Spectral Lines and Bands

The nature of the atomic or molecular structural changes will determine the energy level of the light emitted or absorbed. For changes involving outer shell, i.e., valence electron transitions, $\Delta E$ and, therefore, $v$ and $\lambda$ are in the UV and visible range. Inner-shell electron transitions lead to X-ray radiation. Molecular transitions involving vibrational and rotational modes lead to infrared radiation.

## Emission and Absorption Spectrometry

When matter is excited to high energy levels, by flame, electric arc or spark, it generally decomposes into atomic ions. The dissociation of electrons is reversed as the ion leaves the high-energy zone and results in emission which is characteristic of each ion present. Transitions include those in which electrons drop from outer to inner orbits as well. These phenomena are responsible for the application of **emission spectrometry** to elemental analysis over wide concentration ranges in the steel, aluminum, and other industries as well as a popular technique for environmental and forensic problems.

In or near the high energy zones, ions will trap electrons in reducing zones to form neutral atoms. When light of frequencies corresponding to electronic transitions of the atoms is passed through the atomic cloud, such frequency light will be absorbed by the atoms and excite them. The extent of such absorption is a measure of what is present and in what quantity. This phenomenon was first observed on solar radiation by Fraunhofer, who noticed precisely defined black lines in the solar spectrum. This led to the identification of the then- as-yet undiscovered new element, helium. Since then, **atomic absorption spectrometry (AAS)** has become one of the trace elemental, particularly metal, analytical chemists' favorite techniques.

It is possible to selectively excite molecules without resulting in their decomposition, strangely enough, by using the most-concentrated energy source, electromagnetic energy. The excited molecules or complex (polyatomic) ions, can emit light energy as they return to less excited states in a phenomenon called **luminescence**, of which **fluorescence** is a type.

**Molecular Absorption Spectrometry**, like AAS, exploits the nature and the amount of light absorbed by a sample. Analytical methodology based on these phenomena are very widely used.

## Beer-Lambert Law

Equation 13-1 describes the relation between energy emitted and absorbed for a single particle. The intensity of the energy emitted or absorbed by an ensemble of particles is, of course, proportional to the number of particles per unit area in the light path and also to the probability of the occurrence of the process. This finds expression as the **Beer-Lambert Law**.

$$\log I_o/I \equiv A = abc \tag{13-2}$$

where $I_o$ is the initial radiation intensity and $I$ is the value after passing through a path length, $b$, of the sample in which the analyte has a concentration of $c$. This law applies to a single species (chemical changes may lead to anomalies) for **monochromatic** light over the **entire** range of electromagnetic radiation.

The proportionality constant in Equation 13-2, $a$, called the absorptivity, varies with wavelength and from substance to substance. When concentration is expressed as M, mol/L, and $b$ in cm, then $a$ becomes $\in$, the **molar absorptivity**. From quantum mechanics we learn that the $\in$ value of an absorption band at $\lambda_{max.}$, the $\lambda$ at which $\in$ is maximum, is a measure of the **transition probability**. Thus, the most intensely "colored' substances are those in which a highly probable transition, such

Table 13.1 **Characteristics of Various Electronic Transitions**

| Transition | λ | ∈ | Probability | Selectivity |
|---|---|---|---|---|
| $\sigma - \sigma^*$ | ~ 100 nm | ~$10^4$ | High | Low |
| $d \leftrightarrow \pi$ | 200-600nm | ~$10^4$ | High | High |
| d - d | 200-600nm | $10^0$-$10^2$ | Low | High |
| $\pi \to \pi^*$ | 200-600nm | ~$10^4$ | High | Low |

as a $\pi - \pi^*$, in conjugated, unsaturated, or aromatic compounds or a d - $\pi^*$, charge transfer band as in $MnO_4^-$ ion or $Fe^{3+}$-phenol complex. Transition metal salts are distinctively but weakly colored because the ligand field transition (d - d) are "forbidden", i.e., have a low probability.

An important feature of a spectral band in evaluating its utility for analytical purposes is its selectivity. In the visible and UV part of the spectrum, involving electronic transitions, one is likely to see quite similar spectra for similar compounds. The spectra of aromatic compounds will change somewhat upon substitution which either changes the electron density in, or enhances conjugation of, the overall $\pi$ electron system.

Most metal chelates of aromatic chelating agents, such as 8-quinolinol or dithizone, have spectra which are almost independent of the nature of the metal ion because these spectra are essentially those of the ligand anion. Interesting exceptions to this are found when charge transfer transitions occur, such as with Fe(II)-1,10-phenanthroline, Fe(III)-8-quinolinol, Ni(II)-dithizone chelates, etc., In these cases, high molar absorbances are coupled with high selectivity. In the case of simple transition metal complexes, when d-d transitions are the origin of their spectra, although the molar absorbances are low, the selectivity is quite high. In general, absorption bands in the infrared have lower molar absorptivities but much higher selectivities than those in the visible and UV.

## Spectrophotometric Determinations

_**Single Components**_. Beer's Law can be applied to analytical problems in several ways.

(1)   One way is by the measurement of the absorbance of a single analyte itself or of a highly absorbing product obtained by the addition of a suitable reagent. A calibration curve is constructed by measuring the absorbances at the wavelength of its maximum absorbance, $\nu_{max}$, or at a series of "useful" $\nu$ values of a series of solutions of the analyte at different concentrations. These data are then subjected to linear regression(s) to obtain (a) working curve(s). With currently available instrumentation, e.g., a diode array spectrophotometer, entire spectra can be acquired in a short time($\sim 1$) and with the help of a spreadsheet, the concentration of an unknown can be estimated by averaging over a number of working curves, with a result of higher statistical validity.

(2)   A second way Beer's Law can be applied to analytical problems is by **photometric titration**, which involves monitoring the absorbance of a solution while a suitable reagent is added and plotting the absorbance at a suitable wavelength as a function of the volume reagent solution added.

Unlike titrations previously described in this book, photometric titrations are **linear** since absorbance is directly proportional to concentration. To be sure, we have demonstrated how to treat pH, pM, and pE data, in the context of the Gran method, to obtain linear titrations. Remember two important benefits of linear titrations: (1) it is a practical way to handle dilute solutions and (2) data can be taken more rapidly, using points on either side of the equivalence point. The best line for each of the two sets (before and after the equivalence point) is obtained by linear regression and their intercept used to locate the equivalence point.

The shape of the titration curve, always consisting of linear portions, will depend on whether one monitors the colored analyte or reagent being consumed, a colored complex being formed, or some combination. Even if all the reactants and products absorb at the monitoring wavelength, a titration is still possible if they have different molar absorbances. Further, a photometric titration can be used to determine more than one analyte if each equivalence point results in a characteristic change of absorbance. For instance, a mixture of Cu(II) and Bi(III) can be photometrically titrated with EDTA. Bi(III), which is titrated first, has a colorless complex, so the absorbance will remain constant, if corrected for dilution. The absorbance will rise during the Cu(II) titration because the Cu-EDTA complex absorbs much more strongly than $Cu^{2+}$, and becomes constant when the second equivalence point is reached.

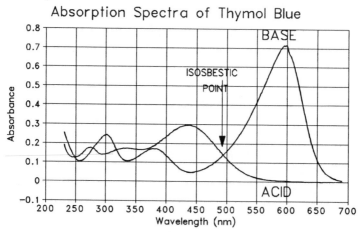

**Figure 13.1 Spectra of Acid/Base Forms of Thymol Blue**

Figure 13.1 illustrates a number of principles in spectrophotometric analysis. The absorbance value of each species is, according to Beer's Law, proportional to its concentration in the solution, and can be seen to change dramatically with the wavelength. The figure also shows how a spectrum can change when the species is protonated. The pi electron distribution of the indicator molecule is sensitive to the incorporation of the proton, which is why it functions as an acid-base indicator.

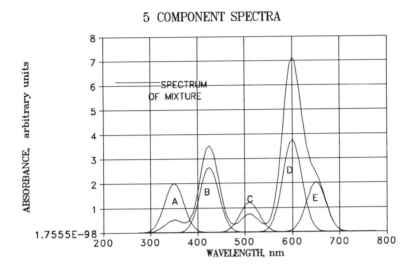

**Figure 13.2 Simulated 5 Component Spectra**

_**Multiple Components**_.  Figure 13.1 also provides a view of how the composition of a multicomponent solution can be determined spectro-photometrically.  In principle, multicomponent spectrophotometric analysis is limited only by the number of wavelengths at which absorbances unique to each analyte present in the mixtures.  This figure presents a special example of binary mixture analysis as the spectrophotometric determination of equilibrium constants possible when the conjugate variables each have characteristically different spectra (See Chapter 18).

As many as six or seven components can be determined.  Naturally, the reliability of both single and, particularly, multiple component determinations would significantly improve if measurements on the sample were carried out on a large number of different wavelengths.  Each such measurement results in an additional independent equation of the variables. The array of these equations can be treated by the spreadsheet Regression function. The improvement will be seen in the smaller standard deviations in the values determined. Naturally, standard spectra representing known concentrations of each of the components of the mixture must be obtained prior to the determinations.

$$y = \frac{1}{\sigma\sqrt{2\pi}} e^{-(x-\mu)^2/2\sigma^2}$$

$$A_A = \frac{C_A \epsilon_A}{\sigma\sqrt{2\pi}} \cdot e^{(\lambda - \lambda_{max_A})^2/2\sigma^2}$$

(12-6),(13-3)

The example illustrated in Figure 13.2 of the determination of a five component mixture was designed with the help of the Gaussian equation 12-6 which was used as an idealized shape of an absorption curve. Simply replacing y by the absorbance, $A_A$, and x and $\mu$ by $\lambda$, results in Equation 13-3 expressing the dependence of $A_A$ on $\lambda$. In column **A**, starting at **A10**, use /EF{/BF; /F} to fill the cells with $\lambda$ values from 280 to 700 nm in steps of 20 nm. Label **A2** to **A6** with $\lambda_{max.A}, .. \lambda_{maxE} =$. In **B2..B6**, we will place values of the $\lambda$ for each of the 5 components, obtained by experiment, literature, or any other suitable means. In **B10..F10**, we will enter the formula for $A_A..A_E$, using either the absolute cell addresses **B2..B6**, or the Block Name method. This will simplify our work in planning other problems and examples. An arbitrary value of 100 was selected for $C_A\epsilon_A$ as well as for all the other components, as was the reasonable value of 30 for $\sigma$ (This should also be entered in terms of an absolute cell address or named block to improve flexibility in problem design). In **G10**, enter a function which is the sum of **B10..F10**, but with each having a multiplier representing a component concentration chosen to simulate a particular mixture of the five components. In Figure 13.2, the concentration multipliers chosen were 0.251, 1.333, 0.606, 1.876, and 1.023 for A to E. You may select any other combination. In constructing the graph, use **A10..A31** for the X series, and **B10..B31, C10..C31, D10..D31, E10..E31, F10..F31**, and **G10..G31**, for the first to the sixth series. You should obtain the same graph as in Figure 13.2.

Now carry out the regression command, /TAR{/AR; /DR} with **B10..F31** as the Independent variable, and **G10..G31** as the Dependent variable. You will note that the X coefficients obtained from the regression analysis match the coefficients exactly. This is, of course, ideal. Try it again, this time multiplying the $A_{mix}$ by a suitable noise function( see Chapter 12), e.g., with $P = 0.01$.

## PROBLEMS

13-1         Table 13.2 lists the molar absorbances for Mn as $MnO_4^-$ and Cr as $Cr_2O_7^{2-}$ at a range of wavelengths as well as an idealized set of absorbances of a binary mixture. Using regression analysis, find the concentration of each of the two metals. Repeat this with the data into which "noise " has been introduced.

**Table 13.2 Molar Absorbances for Mn(VII) and Cr(VI) as $F(\lambda)$**

| $\lambda$, nm | $A_{Cr}$ | $A_{Mn}$ | A(Mix) | A + noise |
|---|---|---|---|---|
| 300 | 1750 | 1750 | 0.6125 | 0.619632 |
| 315 | 1150 | 1800 | 0.565 | 0.5666 |
| 325 | 1400 | 1400 | 0.49 | 0.483557 |
| 350 | 2200 | 1250 | 0.5325 | 0.52561 |
| 375 | 1400 | 600 | 0.29 | 0.290779 |
| 400 | 250 | 150 | 0.0625 | 0.062112 |
| 425 | 300 | 150 | 0.0675 | 0.067057 |
| 450 | 350 | 200 | 0.085 | 0.085067 |
| 460 | 300 | 300 | 0.105 | 0.105417 |
| 500 | 250 | 450 | 0.1375 | 0.135111 |
| 520 | 200 | 1300 | 0.345 | 0.344 |
| 525 | 100 | 1500 | 0.385 | 0.385207 |
| 540 | 90 | 1100 | 0.284 | 0.280181 |
| 550 | 20 | 1300 | 0.327 | 0.327975 |
| 575 | 10 | 450 | 0.1135 | 0.113184 |
| 600 | 0 | 250 | 0.0625 | 0.061589 |

13-2    A calibration curve is to be constructed for the spectrophotometric
        determination of phenol in water using its absorbance at 225nm. The
        following data, presented as concentration(ppm), absorbance pairs, are:
        blank, 0.0.002; 1.0, 0.041; 2.0,  0.083; 3.0, 0.122; 4.0, 0.156; 5.0,
        0.190; 6.0, 0.230; 7.0, 0.265; 8.0, 0.308; 9.0, 0.340; 10.0, 0.378. Is
        the calibration best represented by a linear plot or a higher order plot.
        Use linear regression to help you decide.

# Chapter 14

# Separation Processes

## GENERAL CONCEPTS

One might reasonably expect that as more highly selective and sensitive instrumental methods of detection and determination become available, a truly remarkable phenomenon of 20th century analytical chemistry, that the necessity of incorporating separation steps in an analytical determination would significantly decrease. This is far from true, however. One reason this has not happened is that more stringent requirements--still lower limits of detection, still greater freedom from interferences--are placed on analysts. These requirements arise, for example, by the needs for super-pure substances for high-tech appreciations, by the decreasing strict environmental standards, etc. The domain of trace analysis keeps moving to lower levels with increasing demands of modern material science and with the increasing pressure of environmental concerns.

Not all separations are conducted in the context of trace analysis. Separations on a macro scale are important in many purification processes and in materials recovery, as in hydrometallurgy. It is fortunate that principles of separations are of such broad scope as to apply almost without change to both trace and macro- level separation problems.

Separations may be usefully classified into (a) those involving phase separation, i.e., in transferring a component of interest from one phase to another distinct phase and (b) those in which differential migration within a single phase. A phase consists of homogeneous material which, if condensed (liquid or solid), has discrete boundaries. Substances in the gaseous phase are totally miscible and form a single phase whose boundaries are defined by the containing vessel. Table 14.1 describes types of separation involving various phase pairs. The tabulated separations include both single stage (or batch) and multistage modes of phase contact.

Separation processes not involving phase separation can arise from differential migration of components in a gas or liquid, under gravity fields such as diffusion and ultra centrifugation, or electrical fields, such as electrophoresis. In this chapter, phase separations will be described.

**Table 14 .1 Classification of Types of Phase Separation Processes**

|        | Gas | Liquid | Solid |
|--------|-----|--------|-------|
| **Gas** | (Diffusion - not phase separation) | Distillation GLC | Sublimation GSC |
| **Liquid** | | Solvent Extraction, ion exchange, CCD, LLC, RPLC | Crystallization, precipitation, Adsorption, LC |
| **Solid** | | | Solid-solid Diffusion, D |

GLC = gas-liquid chromatography
GSC = gas-solid chromatography
CCD = counter current distribution
LLC = L-L chromatography
RPLC = reversed phase liquid chromatography
LC = liquid chromatography

## Fundamental Definitions

Without specifying the specific phase pair involved, we can define distribution of an analyte between two phases in terms of the following definitions.

1.  **Distribution Ratio**, D. The distribution ratio is simply the stoichiometric ratio of concentration of a compound in each of the two phases

$$D_A = C_{A(1)}/C_{A(2)} \qquad (14\text{-}1)$$

In many solvent extraction systems, phase (1) and (2) are organic and aqueous phases, respectively. In adsorption, the pair include solid and liquid; in counter-current distribution, the pair refer to immobile and mobile. The distribution ratio D is not to be confused with a distribution constant, a parameter that will be described in later chapters, particularly 14. D, however, is descriptive of a system in equilibrium and its values can be related to the equilibrium constants of all of the processes involved in the particular system.

2.      **Phase Volume Ratio, $R_V$**

$$R_V = V_1/V_2 \qquad (14\text{-}2)$$

where $V_1$ and $V_2$ are the volumes of each phase.

3.      **Distribution Fractions, p and q.** These are the fraction of the component in each of the two phases, i.e., the amount in the phase divided by the total, or the sum of the amounts in the two phases. Remembering that these amount can be described as the product of concentration and volume then gives

$$p - \frac{C_1V_1}{C_1V_1+C_1V_2} \ and \ q - \frac{C_2V_2}{C_1V_1+C_2V_2} \qquad (14\text{-}3)$$

Dividing top and bottom by $C_2V_2$ and incorporating Equations 14-1 and 14-2.

$$p = DR_V/DR_V + 1 \text{ and } q = 1/DR_V + 1 \qquad (14\text{-}4)$$

The **percentage extracted** is simply 100 p.

**Separation Factor, $\alpha$**

As a measure of degree of separation, the ratio of distribution rations, $D_A/D_B$, is termed the separation factor $\alpha$. While it is expected that an increased $\alpha$ signifies an improvement in separation, we must remember that the $D_A$ should be greater than one and $D_B$ should be less. In fact, the best separation to be achieved at a particular $\alpha$ value is obtained at

$$D_A D_B = 1 \qquad (14\text{-}5)$$

**Example 14.1.** Consider a mixture of two solutes A and B initially at the same concentration which are distributing between two phases characterized by $D_A = 100$ and $D_B = 1$ and another equimolar pair, C and D, with distribution ratios, $D_C = 10$ and $D_D = 0.1$. Find the $\alpha$ values and the distribution fractions, p and q, for both pairs.

$$\alpha_{AB} = D_A/D_B = 100/1 \quad \alpha_{CD} = D_C/D_D = 10/0.1 = 100$$

Assuming $R_V = 1$,

$$P_A = 100/100 + 1 = 0.99 \quad P_B = 1/1H = 0.50$$

$$P_C = 10/10 + 1 = 0.91 \quad P_D = 1/10 + 1 = 0.09$$

Thus, although both pairs have the same $\alpha = 100$, the fraction of A in phase 1, 0.99, is only twice that of B, 0.50. The corresponding fractions of C and D are 0.91 and 0.09 which represents a 10-fold enrichment of C over D in phase 1.

To this point, the discussion has been focused on what separation can be achieved with a single stage or batch method. Properly conducted repetitions can greatly increase separation efficiency even when the separation factor, $\alpha$, of neighboring pairs is quite close to unity. When the substance being separated has a low D value, and the corresponding fraction, p, is insufficient for a desired separation, the process may be repeated with a new portion of the second phase until the accumulated amount removed reaches the target level. Consider adding a volume $V_D$ of ethyl ether to an aqueous solution

## Multistage Separations

While single stage processes provide satisfactory solutions to many separation problems, there are at least two conditions that require more.

1.  When the separation factor is close to one, i.e. the distribution ratios of a pair of analytes are too close to obtain complete separation in a single stage separation, then multistage separations, usually in a counter current mode, must be employed (see Chapters 15 and 16).

2.  When the D value of an analyte is so low that a single stage separation would only transfer a small fraction of this analyte, then

the phase containing the analyte could be repeatedly (or continuously) contacted with the second phase.

If only 10% of a substance transfers from a phase in which it is dissolved into an equal volume of a second phase, then its distribution ratio is given by D/D + 1 = 0.1, or D = 0.11. The fraction remaining is:

$$q = 1/D+1 = 0.89$$

If a second transfer occurs, the fraction remaining is 0.89 x 0.89 or $(1/D+1)^2$. After n such transfers, the fraction that remains is

$$\text{Fraction Remaining} = (1/DR_v+1)^n \qquad (14\text{-}6)$$

1.  Separations involving phase transfer: distillation, sublimation, liquid-liquid distribution, gas-solid, and liquid-solid adsorption, both single and multi-stage.

## PROBLEMS

14-1    Ethanol in an aqueous solution whose volume is 150 mL is to be quantitatively removed (at least 99%) by extraction with 10-mL portions of benzene. If the D of ethanol between benzene and water is 0.5, how many such extractions must be made to accomplish the removal?

14-2    In the previous example, the transfer desired was from water to benzene. Now, if the ethanol were initially in the benzene phase, and its removal by extraction by water was desired, how many extractions would be needed to obtain a 99% transfer?

# Chapter 15

# Solvent Extraction Equilibria

---

## SOLVENT EXTRACTION EQUILIBRIA

Liquid-liquid extraction (solvent extraction) procedures have proven very useful in analytical separations. These compare favorably with precipitation procedures because of (a) the virtual absence of phenomena resembling coprecipitation, (b) the applicability to the separation of traces of substances, and (c) the rapidity of separation. Further, extraction procedures are applicable to both organic and inorganic substances.

Like precipitation, solvent extraction involves the removal of a substance from one phase (usually aqueous) into another (usually organic). A study of how the extent of extraction varies with experimental conditions is based on equilibrium considerations. In this chapter as well as in Chapters 16 and 17, the elementary relationships describing phase separation equilibria developed in Chapter 14 will be used.

Here, we will develop relationships describing the empirical measure of extraction, the distribution ratio, D, in terms of all of the relevant equilibrium relationships, for several typical extraction systems. We will see that such relationships can be quite complex. This is not a failure of the Nernst Distribution Law, however. It demonstrates that not only $K_D$, but all the relevant K values and accompanying concentration variables, are important.

For example, the extent of extraction of organic acids and bases will depend on the pH of the aqueous phase.

### Example 15.1

How does benzoic acid (HOBz) extract in a water:i-octanol system at various pH values.

The equilibria to consider are (1) the distribution of HOBz between $H_2O$ and iOcOH:

$$K_D = [HOBz]_o/[HOBz]$$

and (2) the acid dissociation of benzoic acid in water:

$$K_a = [H^+][OBz^-]/[HOBz]$$

The distribution ratio then is

$$D - \frac{[HOBz]_o}{[HOBz] + [OBz^-]}$$

dividing numerator and denominator by [HOBz],

$$D - \frac{K_D}{1 + \dfrac{[OBz^-]}{[HOBz]}} - \frac{K_D}{1 + \dfrac{K_a}{[H^+]}} - \frac{K_D[H^+]}{[H^+] + K_a} - K_D\alpha_o$$

Thus, the distribution ratio of the benzoic acid is seen to be governed by the $K_D$ and $\alpha_o$, the fraction of the analyte representing the neutral, extractable species.

A logD-pH diagram should resemble the log$\alpha$-pH diagrams developed in Chapter 4, with a maximum logD = log$K_D$, i.e., when log$\alpha_o$ = 0. (See Figure 15.1)

A system in which a solute is involved in both phases is illustrated with the distribution of benzoic acid between benzene and water. This system differs from the first one because, in addition to the aqueous dissociation of the acid, it dimerizes (via H-bonding) in the organic phase.

$$2HOBz(o) \rightleftharpoons (HOBz)_2(o)$$

Hence,    $D - \dfrac{[HOBz]_o + 2[(HOBz)_2]_o}{[HOBz] + [OBz^-]}$

Dividing both numerator and denominator by [HOBz], and making the appropriate substitutions:

$$D = \frac{K_D(1 + 2K_p[HOBz]_o)}{1 + K_a/[H^+]}$$

or

$$D = \alpha_o K_D(1 + 2K_p[HOBz]_o)$$

From this expression it may be seen that the distribution ratio of acetic acid depends on the pH of the aqueous phase (via $\alpha_o$ and on the absolute value of the concentration of HOAc present). In general, when the degree of association of a solute in one phase is different from that of the other phase the value of D will vary with the total concentration of the solute present.

The distribution of $I_2$ between water and $CCl_4$ may be described in terms of the simple distribution law, i.e.,

$$D = K_D = [I_2]_o/[I_2]$$

In the presence of KI in the aqueous phase, however, iodine is involved in the reaction:

$$I_2 + I^- \rightleftharpoons I_3^- \quad K = [I_3^-]/[I_2][I^-]$$

Now the total aqueous $I_2 = [I_2] + [I_3^-]$

Therefore  $D = \dfrac{[I_2]_o}{[I_2] + [I_3^-]}$

or,  $D = \dfrac{[I_2]_o/[I_2]}{1 + [I_3^-]/[I_2]}$

Finally,  $D = \dfrac{K_D}{1 + K[I^-]}$

We see that the D will decrease with the $[I^-]$.

From these examples, as well as all the others, we can conclude that, while the expressions for D become complex in the presence of chemical processes other than simple phase distribution, this is not due to a failure in the Nernst Distribution Law, but a natural consequence of accounting for all the participating equilibria.

## ION PAIR EXTRACTION SYSTEMS

Ion-pair formation can also result in an uncharged species having solubility characteristics suitable for their extraction into organic solvents. Reactions involving ion-pair formation are widely used for the extraction of metal ions as well as nonmetallic inorganic and organic ions. Typical counterions used in such extractions are quaternary ammonium ions, $R_4N^+$, where R is an alkyl group of at least four carbon atoms to pair with the desired anion. To extract cations, such anions as $BF_4^-$, $ClO_4^-$, or organic anions such as alkyl sulfates and sulfonates are used. The metal may be incorporated into a very large ion containing bulky organic groups or it may pair (or associate) with an oppositely charged ion which is large. For example, [Fe (phenanthroline)$_3^{2+}$, $2ClO_4^-$], [$(C_4H_9)_4N^+$, $FeCl_4^-$), [$(C_6H_5)_4$ As$^+$, $MnO_4^-$]$^3$ represent extractable ion pairs of analytical utility.

Since many ion-association extractions take place from aqueous solutions of relatively large ionic strengths, a region in which great differences exist between concentrations and activities, it is very difficult to describe quantitatively the behavior of such systems in terms of simple equilibrium expressions. Allowing for the uncertainty in the values of the appropriate conditional formation constants, however, it is possible to derive expressions which are at least qualitatively useful. As an illustration, let us consider the extraction of $Fe^{3+}$ from an HCl solution using a benzene solution of a high molecular weight amine such as tribenzylamine, $(C_6H_5CH_2)_3N$, symbolized by $R_3N$.

$$R_3NH^+, Cl^- \qquad\qquad R_3NH^+, FeCl_4^-$$

ORGANIC

_____

AQUEOUS

$$R_3NH^+, Cl^- \; == \; R_3NH^+ + Cl^-$$
$$R_3NH^+ + FeCl_4^- \; == \; R_3NH^+, FeCl_4^-$$

It has been assumed that sufficient HCl is present in the aqueous phase to permit the neglect of unprotonated tribenzylamine.

$$D_{Fe} = \frac{[R_4N^+, FeCl_4^-]_o}{C_{Fe}} = \frac{K_{D_{Fe}} \cdot K_{IP_{Fe}}}{K_{D_{Cl}} \cdot K_{IP_{Cl}}[Cl^-]}$$

For this system the distribution ratio of iron is given by which formally resembles the expression shown earlier for the distribution of acetic acid between water and benzene.

By the incorporation of the various equilibrium expressions as indicated by the above scheme, the distribution ratio D may be expressed in terms of the chloride ion concentration. With ion association systems, the extracted ion pair-complex can dimerize, particularly in solvents of low dielectric constant. This would result in the dependence of D on the total iron concentration.

## CHELATE EXTRACTION EQUILIBRIA

A large number of chelating agents which may be represented generally as HL (weak monoprotic acids) form uncharged metal chelates which are soluble in organic solvents and are therefore extractable. Such reagents, which include acetylacetone, dimethylglyoxime, and diphenylthiocarbazone, are generally dissolved in organic solvents immiscible with water.

The equilibria involved in a typical chelate extraction are seen in the following scheme.

$$HL \quad ML_n + bB \;=^{Kad}= \; ML_nB_b$$

ORGANIC

_____

AQUEOUS

$$HL =^{Ka}= H^+ + L^-$$

$$M^{n+} + nL^- =^{\beta}_{MLn}= ML_n$$

$$M^{n+} + Y^{i-} == MY^{(i-n)-}$$
(Y is a masking agent, like EDTA)

$$M^{n+} + aX^- == MX_a^{(n-a)+}$$
(X can be OH$^-$, Cl-, SO$_4^{2-}$, NO$_3^-$,etc.)

Other reactions involving the metal ion such as (a) formation of hydroxy complexes, of complexes with other anions present in solution, e.g. $Cl^-$ or $SO_4^{2-} \cdot MCl^{(n-1)+}$, $MCl_2^{(n-2)+}$, and (b) formation of complexes with masking agents, e.g., $CN^-$ or EDTA, e.g., $MY^{(4-n)-}$, are all taken into account by the now familiar $\alpha_M$ introduced in Chapter 5. The expression for D representing the ratio of the total metal ion concentration in each phase is

$$D_M = \frac{K_{D_M} K_a^n \beta_{ML_n}}{K_{D_R^n}} \cdot \frac{([HL]_o^n}{[H^+]^n} \cdot \alpha_M$$

This equation shows, by the absence of any metal concentration terms, that the extraction of a metal is independent of the total amount of metal present and hence may be used for trace and macro levels alike. For a

EXTRACTIONS WITH DITHIZONE
FOR Ag, Cu, Bi, AND Zn

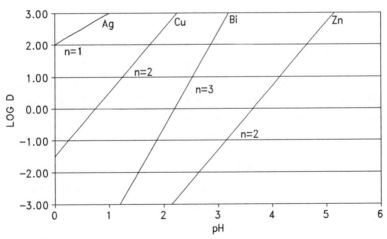

**Figure 15.1  Extraction of Several Dithizonates**

given metal chelate system the extent of extraction increases with the concentration of the ligand in the organic phase and with the pH of the aqueous phase.

The course of solvent extraction of a series of metal ions with a given chelating agent can be conveniently represented in Figure 15.1 where the variation of log D with pH for three metal ions at constant reagent concentration is shown. Each of the plots is seen to have a linear portion whose slope is equal to the charge of the metal ion (strictly speaking, the slope is equal to the number of hydrogen ions released per metal ion in the overall extraction equation). If the log D scale could be extended high enough (with dithiozone chelates this is very difficult because the log $K_D$ is very large), log D becomes constant when $D = K_{DC}$. In this horizontal portion of the curve, the intrinsic solubility characteristics of the chelate (as reflected by $K_{DC}$) rather than the formation equilibria limits the extent of extraction. From the curves it can be seen that the more highly charged the metal cation, the narrower the pH range required for complete extraction. For most analytical purposes, extraction may be considered to be complete at 99% extraction. This corresponds to a value of log $D = 2$ (assuming a phase volume ratio of unity). Since we can usually ignore extractions of less than 1%, the practical limits of extraction of a particular metal are

$$-2 < \log D < +2$$

By examining extraction data for various chelating agents represented in graphical form, such as Figure 15.1, it is possible to predict which metals may be quantitatively separated from each other by one extraction as well as the pH range in which this separation should be carried out. Complete separation will be possible if there is a pH, $pH_A$ where log $D = -2$ for metal II when log $D = +2$ for metal I. The minimum separation between the two curves must be $4/n$.

This relationship does not hold if the two metals under consideration have extraction curves whose slopes are not the same. In general, for a series of extraction curves that may have different slopes, the distance separating each curve has to be described in terms of some arbitrarily chosen value of D as a reference. Usually, the value log $D = 0$ which corresponds to half extraction is chosen for reference and the pH at which this occurs is designated as $pH_{1/2}$.

From the foregoing discussion, it is easy to see how solvent extraction measurements can be used to study various solution equilibria involving metal ions. By determining the variation of D over a variety of reaction conditions, the equilibrium constants of chelate formation, of hydroxy, and other complex formations of quite a number of metal ions have been evaluated.

## PROBLEMS

**15-1**  Show that D ($H_2O:CHCl_3$) for aniline, $C_6H_5NH_2$, a weak base, is a product of its $K_D$ and $\alpha_1$, the fraction representing the neutral, extractable species. Develop a log D-pH diagram.

**15-2**  A solution of 100 mL containing $1 \times 10^{-4}$ each of $Cu^{2+}$, $Ni^{2+}$, and $UO_2^{2+}$ is equilibrated with an equal volume of 0.02 M 8-quinolinol(HQ) in chloroform at a series of pH values in the range of 2 to 10. The distribution constant of HQ is 410, that of each of the $MQ_2$ chelates is assumed to be 850. Using the $pK_a$ values of $H_{2Q+}$ and HQ as 5.0 and 9.90, and the log $\beta$ values are 25.0, 20.0, and 16.5, respectively. Derive the log $D_M$ - pH curves for each of these metal ions. What per cent of $Cu^{2+}$ and $UO_2^{2+}$ are extracted when 2% of $Ni^{2+}$ has been extracted. Repeat this for 2% extraction of $UO_2^{2+}$.

**15-3**  Repeat problem 15.2 when the only change has been the addition of 0.005 M EDTA.

# Chapter 16

# Ion Exchange Equilibria

Naturally occurring minerals called zeolites have long been known to be capable of trading cations in their rather open crystalline lattice for others present in a solution in contact with the zeolite. This is an example of **ion exchange**, which has become a very popular separation process, particularly since, some 55 years ago, synthetic resins which are capable of exchanging cations or anions were developed. These synthetic resins have considerably more exchange capacity than the zeolites, which they almost totally replaced.

The process consists essentially of contacting a solution of an aqueous sample with a given amount of finely divided resin granules, each of which plays a part in the ion-exchange process. In this chapter, we will focus on single-stage, or batch, equilibria. The role of ion exchange in multistage separations will be described in the next chapter. Both single- and multistage ion exchange processes are analytically useful.

Synthetic ion-exchange resins fall into two categories, cationic and anionic exchangers. In each type, the polymeric molecules of the resin contain functional groups, e.g., a sulfonate anion in the cation exchanger or a quaternary ammonium cation in the anion exchanger. Commonly available resins have from 3 to 5 mmol/g of exchangeable groups. These functional groups are associated with oppositely charged ions, or counter ions, that can exchange with other oppositely charged groups present in an aqueous phase.

Ion exchange resins absorb and swell with significant amounts of water, depending on the ionic strength of the solution as well as on the degree of "crosslinking" of the resin polymer. Most are derivatized polystyrene polymers. Polystyrene ion-exchange resins consist of a long aliphatic hydrocarbon backbone to which appropriately substituted phenyl rings are periodically attached. This essentially "stringy" macromolecule is bonded by a crosslinking agent, usually divinylbenzene (DVB) which forms a solid-like, three-dimensional network.

When the percentage of DVB is low, 2 to 4%, the network is open and the extent of swelling with water is high. Resins with 8% DVB are very popular and are structurally stronger but still imbibe water. In its hydrated form, this resin acts like a gel which, by the levels of sulfonate or tetra-alkylammonium ions present, simulate a strong electrolyte of from 3 to 5M ionic strength. Ions exchange and diffuse back and forth. In this sense, ion exchange is quite analogous to liquid-liquid extraction.

The exchange reaction can be described in terms of the following equations.

Cation exchange:

$$RSO_3^-, H^+ + Na^+ \rightleftharpoons RSO_3^-, Na^+ + H^+$$

Anion exchange:

$$R_4N^+, OH^- + Cl^- \rightleftharpoons R_4N^+, Cl^- + OH^-$$

Equilibrium expressions corresponding to these reactions may be written as follows:

$$*K_{ex.1} = \frac{(a_{Na^+})_{resin} \cdot a_{H^+}}{a_{Na^+} \cdot (a_{H^+})_{resin}} \tag{16-1}$$

and

$$*K = \frac{(a_{Cl^-})_{resin} \cdot a_{OH^-}}{a_{Cl^-} \cdot (a_{OH^-})_{resin}} \tag{16-2}$$

In terms of concentration, $K_{ex.1}$ for the $Na^+$-$H^+$ exchange becomes:

$$*K_{ex.1} = \frac{[Na^+]_R[H^+]}{[Na^+][H^+]_R} = K_{ex.1}\frac{\gamma_{Na_R}\gamma_{H^+}}{\gamma_{H_R}\gamma_{Na^+}}$$

where $[Na^+]_R$ refers to the concentration of the cation in the resin phase in millimoles per gram, and $K_{ex.1}$ is the concentration equilibrium constant, determined experimentally.

Values of $K'_{ex.}$ for various cation exchange reactions can be shown to be related to the ratio of activity coefficients in the resin phase by means of the **Donnan theory** developed by describing membrane equilibria. The Donnan theory, based in thermodynamics, states that the activity product of a salt on either side of a membrane to which it is permeable, is identical. We can consider the resin phase as being separated from the aqueous phase by the equivalent of a membrane which permitted the free passage of all ions except those which are chemically bound to the polymer. Applying this to a system consisting of a resin in the hydrogen form in equilibrium with a solution of NaCl, we may write:

$$(a_{Na+} \cdot a_{Cl-})_{resin} = a_{Na+} \cdot a_{Cl-}$$

$$(a_{H+} \cdot a_{Cl+})_{resin} = a_{H+} \cdot a_{Cl-}$$

Dividing one by the other the Cl⁻ terms are eliminated, giving:

$$\frac{a_{NaR}}{a_{HR}} = \frac{a_{Na}}{a_H}$$

Rearranging: $\quad \dfrac{a_{NaR} a_H}{a_{Na} a_{HR}} = 1$

Proving that $\quad {}^*K_{ex} = 1$

Hence, we may rewrite Equation 16-2 as

$$K_{ex} = \frac{[Na]_R[H^+]}{[Na][H]_R} = \frac{\gamma_{Na}\gamma_{HR}}{\gamma_{NaR}\gamma_H}$$

One can interpret this as signifying that ion exchange equilibria depend on the differences in the properties of the ions themselves rather than on any specific interaction with the resin. Ionic activity coefficients of ions of the same charge are very similar to one another, but their activity coefficients in the resin phase are not. Even though it is not possible to determine the coefficients of the cations in the resin phase directly, the resemblance of the resin phase to a highly concentrated electrolyte solution permits the comparison of experimentally determined $^*K'_{ex.}$ values with the activity coefficient ratios of the ions in solutions of high ionic strengths. In a series of exchanges of monovalent cations, the observed order of $^*K'_{ex.}$ values was found to be

$$Cs^+ > Rb^+ > K^+ > Na^+ > H^+ > Li^+$$

Except for the reversal of $H^+$ and $Li^+$, this is the order that is expected on the basis of the activity coefficients of these ions.

$K'_{ex.}$ for a general reaction,

$$nRM_1 + M_2^{n+} \rightleftharpoons R_nM_2 + nM_1^+$$

where R stands for the resin anion and $M_1$ a monovalent cation and $M_2$ an n-valent cation, can be shown to be

$$K'_{ex} = \frac{[MR_n][H]^n}{[M^{n+}][HR]^n} \tag{16-3}$$

Equation 16.3 presents us with an unusual question. How can we define concentration in the resin phase? Traditionally, components in solid phases are described by mole fractions. We will follow a treatment developed by A. Ringbom (*Complexation in Analytical Chemistry*, John Wiley & Sons, New York, 1963) has proven to be practical. This leads to a mixed equilibrium expression in which the $K_{ex}$ is a concentration constant and the concentrations of the ions are expressed by molarity in both the solution phase and the resin phase. In the resin this will be written as millimoles per gram resin, i.e., as if the density of the resin is unity. Table 16.1 lists a series of such exchange constants which are applicable at low loadings such as $(1/n)$ mmol/g where n is the ionic charge. The resins are obtained with from 3 to 5 mmol/g of exchangeable $H^+$. This is referred to as the **capacity** of the resin. In the examples and problems we will assume a resin with $(5/n)$ mmol/g capacity.

## Table 16.1 Exchange Constants for
## Selected Cations on Dowex 50
## (at 8% Crosslinking) in the $H^+$ Form

$$M^{n+} + nH\text{-}R \rightleftharpoons M\text{-}R_n + nH^+$$

| Ion | $K_{ex}$ | Ion | $K_{ex}$ |
|---|---|---|---|
| $Li^+$ | 0.79 | $Ni^{2+}$ | 1.88 |
| $Na^+$ | 1.56 | $Mn^{2+}$ | 2.04 |
| $NH_4^+$ | 2.01 | $Ca^{2+}$ | 3.24 |
| $K^+$ | 2.28 | $Sr^{2+}$ | 5.15 |
| $Ag^+$ | 6.70 | $Pb^{2+}$ | 11.97 |
| $Mg^{2+}$ | 1.32 | $Ba^{2+}$ | 16.16 |
| $Zn^{2+}$ | 1.46 | $Cr^{3+}$ | 8.0 |
| $Co^{2+}$ | 1.72 | $Ce^{3+}$ | 21.95 |
| $Cu^{2+}$ | 1.82 | $La^{3+}$ | 21.95 |
| $Cd^{2+}$ | 1.85 | | |

To obtain a $K_{ex}$ value when the resin is in some other form than the H-form, simply combine constants from Table 16.1 in the usual manner, so as to make the H cancel out.

## Example 16.1  Find $K_{ex}$ for $Ba^{2+}$ when the resin is in the $Li^+$ form

Since the $K_{ex}$ of $Ba^{2+}$ and $Li^+$ for exchange with resin in the $H^+$ form simply multiply the $Li^+$ - $H^+$ equation by 2 and subtract this result from the $Ba^{2+}$ - H equation.  This will cancel out the protons on each side to give

$$Ba^{2+} + 2Li\text{-}R \rightleftharpoons Ba\text{-}R_2 + 2 Li^+$$

The new exchange constant is

$$K_{ex} = K_{ex(Ba)}/K^2_{ex(Li)} = 16.16/(0.79)^2 = 25.89$$

Ion exchange resins loaded with a singly charged ion can readily quantitatively remove small quantities of ions of higher charge and, when conditions are favorable, do a fair job on singly charged ions.

The course of an ion exchange separation can be described in terms of D, the distribution ratio, and the percent E, or percent exchanged.  D will be written as the ratio of ion concentration in the resin phase, millimoles per gram, to the ionic concentration in the aqueous phase, millimole per milliliter (molarity).

## Example 16.2

What percent exchange occurs when 1.0 g of resin in the H-form is equilibrated with

(a)      50 mL of 0.001 M KCl

$$K_{ex} = [K^+]_R[H^+]/[K^+][H^+]_R = 2.28$$

$$D_K \equiv [K^+]_R/K^+] = 2.28 \, ([H^+]_R/[H^+])$$

There are $50 \times 0.001 = 0.05$ mmol $K^+$ which if quantitatively exchanged would give $[H^+]_R = 5 - 0.05 = 4.95$ mmol/g and $[H^+] = 0.001$. Hence,

$$D_K = 2.28 \, (4.95/0.001) = 11286$$

The %$K^+$ exchanged $= 100[K^+]_R w/([K^+]_R w + [K^+]V)$

Dividing both numerator and denominator by $[K^+]$ where w is the weight of resin in grams.

%K exchanged $= 100 \, Dw/(Dw + V) = 99.7\%$

The error in using 4.95000 rather than 4.950015 and 0.001000 instead of 0.001003 is negligible. How could this be totally eliminated?

(b)      How will a change of $[H^+]$ affect the $D_K$? Take 50 mL of a 0.1 M HCl solution containing 0.2 mmol $K^+$ as in (a).

Assuming that the entire amount of $K^+$ will exchange, then

$$D_K = 2.28 \, ([H^+]_R/[H^+]) \approx 2.28 \, (4.8/0.104) = 105.2$$

This is the maximum value of D, and leads to the value of %K exchanged as

$$\%K = 100Dw/(Dw + V) = 100 \times 105/(105 + 50) = 67.7\%$$

Obviously, the assumption of total exchange is a poor one. The exact formula for D is

$$D_K - 2.28 \cdot \frac{\dfrac{4.8 + 0.2D}{D + 50}}{0.104 - \dfrac{0.2D}{50(D + 50)}}$$

This formidable equation yields readily to the method of successive approximations (See Chapter 1), as seen from the following spreadsheet table.

| D ?    | D calc |
|--------|--------|
| 105.   | 105.23 |
| 105.23 | 111.09 |
| 111.09 | 111.10 |
| 111.10 | 111.20 |
| 111.20 | 111.20 |
| 111.20 | 111.21 |
| 111.21 | 111.21 |

In column **A2** is a guess for D. **B2** has the spreadsheet version of the equation above. In cell **A3**, write : +**B2**. Now copy **B2** into **B3**, and do /EC{/BC; /C} of **A3..B9** to obtain series of successive approximations of D. For a D value good to 0.1, the third or fourth approximation would be sufficient. Although the value 105 was taken because it is the solution to the simple equation, we could have started with any number, say 25, and still had sufficiently rapid convergence to come to 111.2 in the same number of approximations.

## Example 16.3

What is the effect of the ionic charge of the cation undergoing exchange?

From Table 16.1, we select $Mn^{2+}$ as a divalent ion whose exchange constant most closely resembles that of $K^+$. How much $Mn^{2+}$ is exchanged when 50 mL of a 0.001 M $MnCl_2$ is equilibrated with 1 g of the H-loaded resin described in Equation 16-2.

$$K_{ex} = [Mn^{2+}]_R[H^+]^2/[Mn^{2+}][H^+]^2_R = 2.04/(1)^2$$

$$D_{Mn} \equiv [Mn^{2+}]_R/[Mn^{2+}] = 2.04 \, ([H^+]_R/[H^+])^2$$

50 X 0.001 = 0.05 mmol $Mn^{2+}$ which would release a maximum of 2 X 0.05 or 0.10 mmol $H^+$ from the resin to give $[H^+]_R$ of 5 - 0.10 or 4.90. $[H^+] = 0.10/50 = 0.002$.

$$D = 2.04 \, (4.90/0.002)^2 = 6.00 \text{ X } 10^6$$

$$\% \text{ Mn exchanged} = 100[Mn]_R \, g/([Mn]_R \, g + [Mn]V)$$

where w is the weight of resin and V the volume of the aqueous solution. Dividing nominator and denominator by $[Mn^{2+}]$.

$$\%\text{Mn exchange} = 100 \, Dw/(Dw + V) \approx 100$$

$$\%\text{Mn still in solution} = 100 \, V/(Dw + V) = 0.8 \text{X} 10^{-3}\%$$

## Use of Complexing Ligands in Ion-Exchange Separations

As we have seen, it is feasible to remove polyvalent cations quantitatively from very dilute solutions by using ion-exchange resins loaded with singly charged cations. We can see from Table 16.1 that ions of the same charge are not so easily separated, however. By using **anionic** ligands to selectively complex one of a pair of metal ions to form an **anionic** complex, we can separate such pairs. It is not advantageous to use neutral ligands for "masking" as one could in extraction separations, because these give cationic complex species that can also be taken up by the cation exchange resin.

If we adopt the criterion for separation of a pair of metal ions that one, M, is at least 99.9% on the resin while the other, M', remains in solution to at least 99.9%, then

$$D_M \geq 10^3(V/w) \text{ and } D_{M'} \leq 10^{-3}(V/w)$$

Using a reasonable 25 to 50 as V/w, a $D_M$ value of at least $10^{4.4}$ to $10^{4.7}$ is needed for complete uptake, while values of $10^{-1.3}$ to $10^{-1.6}$ are maximum for the metal to remain in solution.

To recognize the influence of complexation, the exchange equilibrium expression can be changed simply by using $C_M = \alpha_M[M^{n+}]$. Hence, the conditional exchange constant, $D'_M$, is written (for a resin in the $Na^+$-form).

$$D_M' = [M^{n+}]_R/C_M = K_{ex}([Na]_R/[Na^+])^n \, \alpha_M$$

If $D_M$ is approximately $10^{4.5}$ then in order for $D'_M$ to be $10^{-1.5}$ so as to prevent uptake of the metal by the resin, $\alpha_M$ should be a maximum of $10^{-6}$. Assuming that a 0.01 M solution of the complexing agent Y, is used then $\beta'_{MY}$ should have a minimum value of $10^8$ to be effective. For this strategy to be useful, the other metal should not form any complexes. At least, by controlling experimental conditions, e.g., by pH control, the complexing agent should react with only one of the metal ions. For example, $Cl^-$ forms anionic complexes with $Zn^{2+}$ and $Cd^{2+}$ but not with $Mg^{2+}$ or $Ca^{2+}$; EDTA forms anionic complexes with all the metal ions first cited. The $MgY^{2-}$ and $CaY^{2-}$ require much higher pH than do the other ions, however. Hence, a pH range in which $\alpha_{Zn}$ is $\leq 10^{-6}$ and $\alpha_{Mg} \sim 1$ should be possible to find.

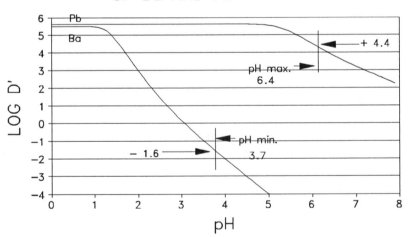

Figure 16.1 Use of Complexation to Enhance Ion Exchange Separations of Metal Ions

## Example 16.4

Can trace levels ($< 10^{-4}$ M) of $Pb^{2+}$ and $Ba^{2+}$ be separated by a batch ion exchange procedure?  Assume that 50 mL of solution which has 0.01 M $Na_2H_2Y$ (EDTA) is to be equilibrated with 2 g Dowex 50 in the $Na^+$ form. Is there a pH range in which the $Pb^{2+}$ will remain in the solution and the $Ba^{2+}$ is taken up by the resin?

The criteria for this to occur is $D'_{Ba} \geq 10^{4.4}$ and $D'_{Pb} \leq 15^{1.6}$.  The exchange equilibrium expressions are:

$$D'_{Ba} = \{16.16/(1.56)^2\}([Na]^2_R/[Na]^2)\alpha_{Ba} = 4.14 \times 10^5\ \alpha_{Ba}$$

$$D'_{Pb} = \{11.97/(1.56)^2\}([Na]^2_R/[Na]^2)\alpha_{Pb} = 3.08 \times 10^5\ \alpha_{Ba}$$

with $[Na^+]_R = 5$ and $[Na^+]$ taken as 0.02.  We will use values of $pK_a(EDTA) = 2.0, 2.7, 6.13$, and $10.33$ and $\log \beta_{PbY} = 18$, $\log \beta_{BaY} = 7.8$.  $\alpha$ varies with pH according to

$$\alpha_{Ba} = 1/(1 + \beta_{BRY}\alpha_Y C_Y) \text{ and } \alpha_{Pb} = 1/(1 + \beta_{PbY}\alpha_Y C_Y)$$

$$\alpha_Y = K_1K_2K_3K_4/(H^4 + K_1H^3 + K_1K_2H^2 + K_1K_2K_3K_4)$$

Prepare a spreadsheet with pH values from 0 to 7 in steps of 0.1 in column **A**, and successive columns with $[H^+]$, $\alpha_Y$, $\alpha_{Pb}$, $\alpha_{Ba}$, $D'_{Pb}$, $D'_{Ba}$, $\log d'_{Pb}$ and $\log D'_{Ba}$ .  From this, develop the graph of $\log D'_{Pb}$ and $\log D'_{Ba}$ vs. pH as shown in Figure 16.1.  From this we see that the desired pH range is 3.7 to 6.4 .  How will alteration of $C_Y$, $[Na^+]$ alter this range? In a hypothetical divalent metal ion pair, what are acceptable values for $D_M$, $D_{M2}$, $\beta_{M1Y}$, and $\beta_{M2Y}$, for the batch ion exchange separation to be feasible? For a tervalent metal ion pair?

## PROBLEMS

**16-1**   Adapt example 16.2 to cases where the size of the sample is higher, e.g., 0.2 to 0.5 mmol, the aqueous concentration of the ion loaded on the resin is higher than that in 16.1b, for mono-, di-, and tervalent cations.

**16-2**   Is it possible to quantitatively separate $Ni^{2+}$ and $Mn^{2+}$ present at $< 10^{-4}$ M levels in 50 mL of a solution containing 0.01 M EDTA using 2 g of a Na-loaded resin? Assume $[Na^+] = 0.03$.

**16-3**   Anion exchange resins have bonded quaternary ammonium ion functional groups ($\sim 3$ mmol/g resin) which are ion paired to exchangeable anions.  For example, the reaction of a Cl-loaded resin with some singly charged anion, X, is

$$R\text{-}Cl + X^- = R\text{-}X + Cl^-$$

The values of the $K_{ex}$ for $X^- = F^-$, $OH^-$, $Br^-$, $I^-$, and $NO_3^-$ are 0.09, 0.09, 2.8, 8.7, 3.8.

Calculate the percent exchange from the equilibration of 30 mL of 0.0002 M NaX with 1 g of the Cl-loaded resin with each of the anions listed.

# Chapter 17

# Multistage Separation Processes

A process designed to separate a pair of substances that results in a separation ratio, $\alpha$, of less than $10^4$ will not adequately separate the pair by a single equilibration. The best remedy is to adjust the chemical parameters or change the process altogether (different reagents, etc.) to obtain the desired $\alpha$. This is not always easy to do. No sooner than chemists solve one set of very difficult separation problems, but they are presented with an even more difficult set! A more general strategy, in the absence of a suitably selective process, is to apply the best available process in a multistage mode. Using this strategy, separations of pairs whose $\alpha$ values are very close to unity have been readily achieved with some of the techniques described here.

This chapter is devoted to detailing the consequences of applying the phase separation principles, previously described in their single stage or batch modes, in a multistage, **countercurrent** manner. The term, countercurrent, refers to the manner in which two phases contact one another, namely, by streaming past one another, flowing in opposite directions. This is also descriptive of systems in which only one phase, called the **mobile phase**, flows past the other, **immobile phase**. In the examples cited here, one phase is fluid, gas or liquid, and mobile. The other, either bulk liquid, liquid supported on a solid surface, or solid remains immobilized. The solutes, introduced with the mobile fluid phase will distribute (or, to use an equivalent British term, partition) between the mobile and immobile phases in a manner depending on their D values. With different solutes, therefore, the migration through the series of stages will proceed to different extents. It is this differential migration that brings about the separation.

As will be seen, there are many variations of these processes which may differ in both mechanisms of phase distribution and operational ways. Nevertheless, they share a number of general characteristics which enable us to describe them in a similar manner. We will start with the countercurrent distribution (CCD), developed by Lyman Craig in the 1930s primarily to help clarify the mechanism of liquid chromatography. Interestingly enough, CCD proved to be a useful multistage separation process in its own right and we will examine it here. Next, we will describe the general aspects of in the factors determining the distribution ratio, D. Thus, what is described here is applicable with little or no change to ion exchange, adsorption chromatography, gas chromatography, etc.

# COUNTERCURRENT DISTRIBUTION (CCD)

For about 15 to 20 years following Craig's development of CCD, the technique found broadly based application to separations of chemically delicate substances that might undergo molecular rearrangements or other decomposition in the presence of the typically surface active adsorbents used in chromatography. In addition, the relatively high capacity of CCD permitted a degree of scaleup that was otherwise difficult to attain. Biochemists interested in enzyme and protein separations and organic chemists found CCD attractive despite the complex and rather bulky (room size in some instances) apparatus required.

In recent years, except for large-scale processes such as in hydrometallurgy, CCD use experienced a decline that ended about 10 years ago with the introduction of centrifugally driven and compact variations, called Planetary Centrifugal Chromatography, invented by Ito and Centrifugal Partition Chromatography (CPC), produced by the Sanki Co., Japan. Despite the incorporation of the term "chromatography" in the name, a move perhaps designed to enhance the commercial acceptance of the technique, these are variations of CCD, and explicable by the same mathematical treatment.

These techniques were introduced to assist biotechnologists with separations on both analytical and preparative scales, but have recently been shown to apply equally well to inorganic and metal ion separations. These techniques can provide separations of thousands of stages with apparatus that does not fill a large room, as did the Craig apparatus. The Sanki instrument is quite compact, occupying a space of about a 2-ft cube, not counting the detector and other accessory equipment.

With CCD, as with liquid partition chromatography, it is possible to select either the aqueous or the organic phase as the mobile phase. The mathematical treatment for each of these is quite similar, requiring merely a switching of **p**, the fraction extracted with **q**, the fraction remaining in the aqueous phase. The chromatographic process in which the aqueous or polar phase is mobile is the considered "normal". When the organic phase is mobile, the term, "reversed phase" chromatography. In our treatment of CCD, we have elected to have the aqueous phase as the mobile one, so as to parallel the case of normal chromatography.

Suppose we have a series of separatory funnels, each of which contains 50 mL of an aqueous phase suitably conditioned (buffered, ionic strength adjusted, etc.). Into the first of these funnels, which we will call funnel **0**, is placed 50 mL of a benzene ($C_6H_6$) solution of an analyte A whose distribution ratio is $D_A$. Funnel **0** is shaken to equilibrate phases. The fraction of A contained in the aqueous phase, called **q** will be given by $(1/(D_A+1)) = q$, while the fraction of A in the $C_6H_6$ phase, $p = D_A/(D_A+1)$ (See Figure 14.3). When the lower, aqueous, phase of funnel **0** is transferred to the next funnel (**1**), the first stage is complete. At this point, the analyte is contained in two funnels (0th and 1st). In **0**, the fraction of A is **p**; in **1**, the fraction is **q**. This designates the **entire** contents of the funnel, including both phases.

The second stage is begun by introducing another, analyte-free, 50-ml portion of aqueous phase into the 0th funnel, and shaking both funnels to equilibrate. With $D_A$ constant, the fraction of A in the aqueous phase in 0th funnel is **pq** while in the upper phase, the fraction is $p^2$. In funnel **1**, which contained **q** before equilibration, the aqueous phase has $q^2$ in the water phase and **pq** in the upper phase. Just as in the last stage, the lower, aqueous phases are transferred to the next, third funnel and the second stage is complete. The contents of funnels 0, 1, and 2 are $p^2$, **2pq**, and $q^2$, respectively. This distribution is described by $(q + p)^2$. In Table 17.1, CCD involving a lower aqueous phase is described for three stages. If there was a fourth stage, the distribution would be represented by $(q + p)^4$. Notice that the analyte is distributed over N + 1 funnels (N = number of stages), i.e., after the 3rd stage, 4 funnels; after the 4th, 5 funnels, and so forth. For N stages, the distribution is given by $(q + p)^N$.

## Table 17.1  Three Stage Countercurrent
## Distribution:  Aqueous Phase Heavier

| | | Upper | Lower | | Equation |
|---|---|---|---|---|---|
| 0th Stage | Funnel 0 | p | q | $= p + q = 1$ | $(p+q)^0$ |
| Transfer:1st stage | | | TOTALS | | $(p+q)^1$ |
| | Funnel 0 | | p | | |
| | Funnel 1 | | q | | |
| Equilibrate: | | | | | |
| | Funnel 0 | $p^2$ | pq | $=p^2 + pq$ | |
| | Funnel 1 | pq | $q^2$ | $= pq + q^2$ | |
| Transfer:2nd stage | | | TOTALS | | $(p+q)^2$ |
| | Funnel 0 | | $p^2$ | | |
| | Funnel 1 | | 2pq | | |
| | Funnel 2 | | $q^2$ | | |
| Equilibrate: | | | | | |
| | Funnel 0 | $p^3$ | $p^2 q$ | $= p^3 + p^2q$ | |
| | Funnel 1 | $2p^2q$ | $2pq^2$ | $=2p^2 q + 2pq^2$ | |
| | Funnel 2 | $pq^2$ | $q^3$ | $= pq^2 + q^3$ | |
| Transfer: 3rd Stage | | | TOTALS | | $(p+q)^3$ |
| | Funnel 0 | | $p^3$ | | |
| | Funnel 1 | | $3p^2q$ | | |
| | Funnel 2 | | $3pq^2$ | | |
| | Funnel 3 | | $q^3$ | | |

It is interesting to remember that the general expression for each term in the **binomial expression** was derived to describe the probability of a coin landing r, heads, and N-r, tails, when tossed N times. This is:

$$P_{N,r} = \frac{N!}{r!(N-r)!} \cdot p^{(N-r)} q^r \qquad (17\text{-}1)$$

where $P_{N,r}$ is the fraction of the total analyte A in the rth stage in an N-stage distribution. When the organic solvent is the phase being transferred at each stage (the mobile phase) rather than the aqueous phase then, while the coefficients in Equation 17-1 would be the same, the pq product would be $p^r q^{(N-r)}$.

Figures 17.1 to 17.3 describe CCD separations in which the aqueous phase is the heavier phase, the one which is transferred successively. The effect of increasing the number of stages on separation efficiency is self-evident. The calculations for these figures are based on Equations 17-1 and 17-2. The appearance of the CCD strongly resembles the bands representing the separation of three analytes (with $D_A$, $D_B$, and $D_C$) seen in chromatograms.

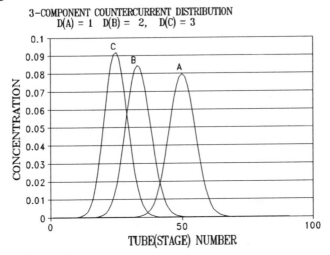

Figure 17.1  100 Stage CCD

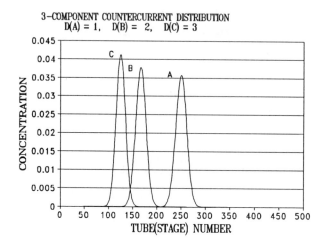

**Figure 17.2  500 Stage CCD**

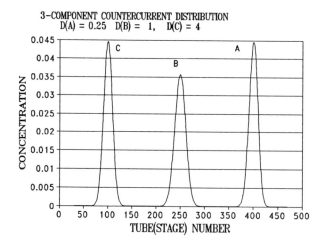

**Figure 17.3 Another 500-Stage CCD**

There is a very important difference, however. Notice, particularly, that the width of the three bands in any of the three figures is independent of the D values. You see, in using the binomial expansion to represent the results of a multistage separation process, only the role played by equilibrium considerations was considered. To obtain results that match experimental reality, however, this must be amended to include the effects of diffusion and other related nonequilibrium processes. This will be dealt with in the next section, which concerns chromatography.

## Spreadsheet Generation of CCD Diagrams

The figures were constructed with the aid of a spreadsheet in which values of **r** are listed in column **A** (starting with **A3**) by /EF{/BF; /F} from O to N (100, 500, or 1000), column **B** (at **B3**) is used for **N - r** using the formula, +**N-A2**. The quantity, N!/(r!(N - r)!), can be obtained without first calculating the values of each factorial separately. In **C3** write **$B$1** (absolute address for N, the number of stages). In **C4**, the formula, +**C3*B3/A4**. In **C5**, write +**C4*B4/A5**, then /EC{/BC; /C} the rest of the column. Here are the equations for the first three values. Can you verify them as well as try to write expressions for the next few terms?

$$N!/r!(N-r)!$$

$$\begin{array}{ll} If\ r-1 :\ -\ \ -N \\ r\ -\ 2 :\ -\ -\ N(N\ -\ 1)/2! \\ r\ -\ 3 :\ -\ -N(N\ -\ 1)(N\ -\ 2)/3! \end{array} \qquad (17\text{-}2)$$

To continue the construction of the spreadsheet, Column **D**, has the formula, +**($G$1*$B$2/($G$1*$B$2 + 1))^$B4**. The absolute cell addresses, **$G$1, $B$2** represent $D_A$, the distribution ratio of the component A, and the phase volume ratio, $R_v$, respectively. The formula itself is one first introduced as Equation 14-4 $p_A$. Similarly, +**(1/($G$1*$B$2 + 1))^$B4**, placed in the next column, represents $q_A$. Now, in column **F**, the formula representing the concentration of A for each stage **r**, write +

**C4\*D4\*F4**, the formula for the proper term of the binomial expansion.[1] This process is repeated for all the other components in the mixture, using appropriate absolute cell addresses for the D values.

# TYPES OF CHROMATOGRAPHY

Although there is a type of chromatography in which the column is simply a hollow tube (**capillary** chromatography), in most cases it is a cylindrical tube packed uniformly with particles of essentially uniform, small (5 to 20 microns) size of a substance capable of selective affinity with various components of a sample. In **adsorption** chromatography, the column is packed with a possibly porous solid with a highly active surface, either polar as with $SiO_2$ or $Al_2O_3$, or nonpolar as with active carbon.

In **partition** chromatography, the particles are either a relatively inert carrier for a dispersed liquid film (either polar or, in **reversed phase**, nonpolar), or a derivatized solid bearing a chemically bonded, generally nonpolar, organic side chain. With the latter type, the organic film is considered to have liquid-like character. With **ion exchange** chromatography, the particles represent a quasi-solid polymeric phase in which regular and frequently spaced functional groups such as $-SO_3^-,H^+$ or $-NR_4^+,X^-$ groups and their associated oppositely charged ions can exchange with ions in a mobile, usually aqueous phase. **Size exclusion** chromatography, as its name implies, relies on pore size of the packing employed to differentiate between substances, frequently polymeric materials. The term **elution** chromatography represents a technique not for conducting chromatography, rather to the mechanism underlying the separation, and applies to most systems. Alternates to elution are no longer in common use.

---

[1]It is sometimes useful to be able to get a satisfactory approximation of a factorial number. This may be obtained by Stirling's approximation

$$N! - \sqrt{2\pi N}(N/e)^N$$

$$or \quad \ln_e N! - 0.5\ln_e 2\pi N + N(\ln_e N - 1)$$

(17-3)

which gives a very good fit, particularly if $N \geq 6$, the value at which the error caused by the approximation is $\sim 1\%$

# LIQUID PARTITION CHROMATOGRAPHY

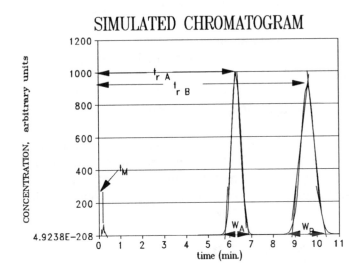

**Figure 17.4 Characteristics of a Chromatogram**

Let us now describe liquid partition chromatography, using the appropriate parts of the development of CCD. A column is packed with a finely divided, uniformly sized solid material whose surface is either derivatized so that it contains a chemically bonded phase, usually organic, or a thin film coating of a liquid in which the mobile phase liquid is very insoluble. The column is first conditioned by equilibrating it with the solvent that is to be used as **eluent**. After a sample solution containing the analytes is placed on the column, the eluent is added. The eluent flows through the column at a controlled (usually constant) flow rate, $\underline{u}$, mL/min until all of the components of the analyte, each at its own characteristic volume called the **retention volume, $V_R$**, have passed through the column. The components of the sample undergo multistage distribution as they pass through the column just as they would in CCD. Each component appears in a zone or band whose concentration profile closely resembles a Gaussian curve, provided the D value remain the same (Figure 17.4). $V_R$ is measured at the maximum concentration at the center of the peak. The retention time, $t_R$, of a particular component is simply $V_R/u$. If a component does not distribute into the immobile phase and, therefore, is not retained, its $V_R$ is called $V_M$, the column **dead, or interstitial volume**.

## Location and Width of Chromatographic Band

**Location of Band.** The value of $V_R$ or $t_R$ is related to the distribution ratio $D = C_A(i)/C_A(m)$ where i and m refer to immobile and mobile phase values, respectively. To obtain a useful quantity called the **capacity factor, k′**, which is the ratio of the number of moles of solute in the stationary phase to the number in the mobile phase

$$k_A' = C_A(i) \times V_i / C_A(m) \times V_m = D_A V_i$$

The quantity $k_A'$ transforms into $p_A$, the fraction of A in the immobile phase by $p_A = k_A'/(k_A' + 1)$. Similarly, the fraction of a solute in the mobile phase, q, is $1/(k' + 1)$. The larger the q value, the smaller the eluent volume needed to sweep a component through the column. Thus,

$$V_R - \frac{V_m}{q} \quad or \quad V_R - V_m(1 + k') - V_m + V_i D \qquad (17\text{-}4)$$

This equation can be modified to emphasize the difference in components by subtracting $V_m$ from $V_R$, to obtain the **net retention volume, $V_N = D V_i$**, or **net retention time, $t_{RN} = V_N/u$**.

**Width of Band.** A vital characteristic of a chromatographic column is its efficiency, commonly expressed by its **number of theoretical plates, N.** N can be calculated from a single band, and is

$$N = t_R^2/\sigma^2$$

where $\sigma$ is standard deviation measured in units of time. The shape of the chromatogram can be represented by a Gaussian curve (see Chapter 12) centered at $t = t_R$ and whose width, W, at half the peak height is $2\sigma$, and at baseline $W = 4\sigma$. Hence,

$$N = 16(t_R/W)^2$$

Another method for approximating N is to determine $W_{1/2}$, the width of a peak at half its maximum height. Then N is

$$N = 5.54 \, (t_R/W_{1/2})^2$$

The value of N calculated from one peak in a chromatogram should be the same for each of the other peaks. Another way to characterize the efficiency of a column is to calculate the quantity N/L, called **H**, the height equivalent per equivalent plate (L is the length of the column). N and H will depend on the number of equilibrium stages in the process, as we have seen above in our discussion of CCD. The actual band widths observed are always larger than ideal because of the nonequilibrium but reproducible processes accompanying mass transfer, such as diffusion of various sorts.

## Simulation of Chromatogram with a Spreadsheet

To construct a simulated chromatogram using the spreadsheet is relatively simple and straightforward when you remember how Equation 12-6 was entered into the spreadsheet to produce the Gaussian curve in chapter 12. We have made use of this already in Chapter 13 when we approximated the shape of an absorption spectrum by a Gaussian curve. In this application of the Gaussian curve, the situation is only slightly more complicated because the width of the curve, as measured by $\sigma$, will increase with the retention volume of the component, on a column of given efficiency, i.e., a given number of theoretical plates. This is taken care of by the definition of $\sigma$ in the equation below. Here, we recognize that the function, y, is multiplied by an amount in micromoles per minute, $\mu$ changes to $t_R$, x to t, $\sigma$ to $t_R/\sqrt{N}$ , etc. to produce

$$y = \frac{1}{\sigma\sqrt{2\pi}} e^{-(x-\mu)^2/2\sigma^2}$$

$$C = \frac{\$MOL_A}{\$SIG_A * \sqrt{2\pi}} e^{(t - \$TR_A)^2/2\$SIG_A^2} \qquad (12\text{-}6), (17\text{-}5)$$

$$\text{where } \$TR_A = \$V*(1 + \$D_A*\$R_v)/\$U$$
$$\text{and } \$SIG_A = \$TR_A/\sqrt{\$N}$$

which is used as the basis for the spreadsheet development.

The $ in front of various symbols in Equation 17-5 indicates that these are **block names**. Earlier we used absolute cell addresses in our spreadsheet formulas to enable us to use a single spreadsheet to serve for a large number of problems. We have used **named blocks** earlier, but it is good to repeat this for the kind of problem we are now considering because of the large number of adjustable parameters employed. By naming blocks, we can make complex equations having many adjustable-parameters much easier to use. For example, by naming the cell that contains the number of theoretical plates, N, then $N can be entered in the formula. This is much easier to remember than say, $B$2, as the cell containing our value for N, and far less subject to mistake.

Other block names used in Equation 17-6, are tabulated here:

| Symbol | Represents: | Symbol | Represents: |
|--------|-------------|--------|-------------|
| N | No. plates | D_A | $^*D_A$ |
| L | Col. length | MOL_A | $^*\mu moles_A$ |
| A | Col. X-area | TR_A | ret. time$_A$ |
| V | $=\$L*\$A$ | SIG_A | $=\$TR_A/\sqrt{\$N}$ |
| u | Flow rate | R_v | $=V_i/V_m$ |

$^*$ Additional symbols representing components B, C, D, E, and F were also provided block names, a total of 30, including those tabulated above.

Set aside at least 60 cells, then, to accommodate not only the blocks but (prefaced by [']), their meanings. A convenient location would be the first 7 to 8 rows in columns **A** to **G**.

Now, starting in column A, say **A10**, use \EF{/BF; /F} to enter a series of times, in 0.1-min. intervals from about 2 to 15, and enter formulas for $C_A$, $C_B$, etc. in successive columns using the formula in Equation 17-5. A seventh column can be written which is the sum of all the C values at each time. As seen in the figures, curves for five of the six components are shown, as well as the envelope of all six present. Mixtures having more than six components can also be solved this way, even though the limitation of the graphics restricts us to a total of six separate curves, so a selection must be made for viewing.

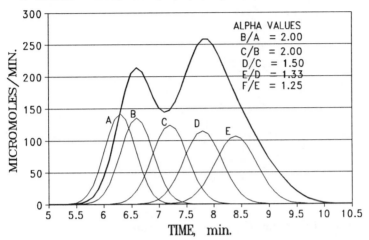

**Figure 17.5 Simulated  500-Plate Chromatogram**

In Figure 17.5, in which 6 components present in equimolar amounts (100μmol) are separated on a 500-plate column.  Note that of the five separate components shown, the height decreases and the width increases with increasing D.  The area under each curve, however, remains constant, demonstrating the advantage of using peak area rather than peak height in analysis. The envelope, as overall chromatogram, shows only two broad and overlapping peaks.    Figure 17.6 illustrates the appearances of chromatograms of mixtures of varying amounts of each of the components and with increasing column efficiency.

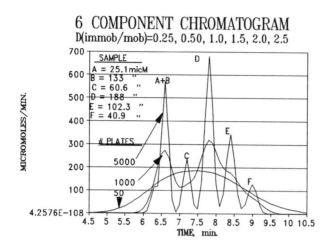

## 6 COMPONENT CHROMATOGRAM
D(immob/mob)=0.25, 0.50, 1.0, 1.5, 2.0, 2.5

**Figure 17.6 Effect of Column Efficiency on Separation**

## Factors Affecting Column Efficiency

As briefly mentioned earlier, multistage separation processes involve not only equilibrium considerations, but kinetic factors as well. The efficiency of a chromatographic process is subject to a number of these factors, such as longitudinal diffusion (B), eddy diffusion (A), mass transfer in the immobile ($C_i$), and mobile ($C_m$) phases. A useful relation, proposed by van Deemter and subsequently elaborated by many others, sums all of these factors

$$L/N = H = B/u + +(1/A +1/C_mu) + C_iu$$

where **H** is height equivalent per theoretical plate and **u** is the flow rate of the mobile phase  The quantity **B** is proportional to $D_m$, mobile phase diffusion coefficient while $C_i$ and $C_m$ are inversely proportional to $D_i$ and $D_m$ for the stationary and mobile phases, respectively.

## Selectivity Factor

The **selectivity factor** $\alpha$ of a column for two species A and B (see Figure 17.5) is defined as

$$\alpha = (t_R)_B - t_M)/(t_R)_A - t_M)$$

or

$$\alpha = D_B/D_A = k'_B/k'_A$$

The selectivity factors for closely related organic compounds are generally relatively low, particularly in systems where the major differences between components are based on very small non-specific interactions, such as Van der Waals forces. If the pair to be separated have acid/base or other complexing capabilities, then manipulation of the experimental environment can increase $\alpha$ values. Such considerations have already been discussed for single stage separations as in Chapters 15 and 16.

## Column Resolution

The **resolution, R**, of a column provides a quantitative measure of its ability to separate two analytes. The significance of this term may be seen by examination of Figure 17.6 which consists of chromatograms for six species on three columns of different efficiencies and therefore, of different resolving powers. The resolution of each column is defined as

$$R_s = \frac{2\Delta t_R}{W_A + W_B} = \frac{2[(t_R)_B - (t_R)_A]}{W_A + W_B} \tag{17-6}$$

A similar equation is readily obtained from this, relating the resolution of a column to its N as well as to the k' and $\alpha$ values of a pair of solutes on the column.

$$R_s = \frac{\sqrt{N}}{4}\left(a - \frac{1}{a}\right)\left(\frac{k_B}{1 + k_B}\right) \tag{17-7}$$

where $k'_B$ is the capacity factor of the slower-moving species. By rearranging this equation, we find the number of plates needed to obtain a given resolution:

$$N = 16R_s^2\left(\frac{\alpha}{\alpha-1}\right)^2\left(\frac{1+k'_B}{k'_B}\right)^2 \qquad (17\text{-}8)$$

The time $(t_R)_B$ required to elute the two species in Figure 17.6 with a resolution of $\mathbf{R_s}$ is:

$$(t_R)_B = \frac{16R_s^2 H}{u}\left(\frac{\alpha}{\alpha-1}\right)^2\frac{(1+k'_B)^3}{(k'_B)^2} \qquad (17\text{-}9)$$

where u is the linear rate of movement of the mobile phase.

## Gradient Elution

Throughout the discussion of chromatography, we have focused on pairs of analytes that were difficult to separate and the need for columns of high efficiency. It is important to remember that in multicomponent mixtures, some analytes are not so difficult to separate. In such cases, another principle must be considered, namely, the time required to move all the components through the column. Further, the longer the residence time, the more diffusion effects will spread the band out. In a mixture in which two adjacent components have reasonably different $k'$ values, then the eluent might be altered when the first component clears the column to reduce the $k'$ of the second. This can be done by continuous or concrete step **gradients**.

Systems that are especially amenable to this technique are those involving chemical interactions. Using the process of solvent extraction as a reasonable model, we can alter the value of D of an analyte two ways. First, D for any substance depends on the organic solvent, because $K_D$ will change. Second, the value of D when chemical interaction occurs in either aqueous or organic phase, can change in very dramatic fashion.

Thus, D of propanol will change when the organic solvent changes from, for example, heptane to chloroform. But if the analyte is propanoic acid, then D in a given solvent pair is not constant but depends on the pH of the aqueous phase, i.e., $D = [H^+]K_D/([H^+] + K_a)$. Controlling the pH in a mixture of solutes will enhance differences due to $K_D$, particularly if there are significant differences in $K_a$. Similarly, when metal complexation can occur in the aqueous phase, control of ligand concentration either directly or, if the ligand is a Brønsted base, by pH control, can effect great changes in the D values of the metal ions involved.

A special case of gradient elution, called **column washing**, can be used. Conditions should be adjusted so that the D of the first analyte is sufficiently small for $V_R$ to be less than half the column volume and that for the second to be large enough for $V_R$ to be greater than five column volumes. If these conditions apply, the first analyte can be quantitatively "washed" out, without too much concern about flow rate or other parameters that optimize column efficiency, without any danger that the second will be removed from the column. When the first analyte is removed, then the eluent is altered so that the D of the second analyte becomes very small, and this analyte, too, is readily washed out.

Column washing has been used in ion-exchange chromatography in conjunction with a complexing agent that results in a metal complex of opposite charge.

## PROBLEMS

17-1    For a two component mixtures having $\alpha$ values of 1.10, how many plates would be required to obtain a baseline separation? Use a spreadsheet to find your answer. Repeat with other small values of $\alpha$.

17-2    Consider the case of a pair of weak organic acids with $K_D$ values of 3.0 and 3.5, whose $pK_a$ values are 4.2 and 4.8, respectively. What are the capacity factors of these acids using an aqueous eluent at pH 1? At pH 4.4? At pH 7.0? How many plates would be needed to separate the acid which elutes first before 0.5% of the second elutes at each of these pH values? Is a gradient elution scheme feasible here?

17-3    A liquid chromatographic column is 20 cm long, and whose mobil phase and stationary phase volumes are 1.20 mL and 0,15 mL, respectively. Using a flow rate of 0.5 mL/min, the retention time of the solvent is 2.5 min and that of analyte A is 6.7 min. If the width of the A band at its base is 0.41 min, find the capacity factor and distribution ratio of A and the number of theoretical plates in the column. How big a D value must a second analyte have to give 'baseline' separation( less than 1% overlap) from A in the chromatogram of this mixture? How would this change if a column having 4000 theoretical plates were used?

# Chapter 18

# Kinetic Methods of Analysis

Most of the analytical methods discussed before now involve selectivity based on appropriate differences in equilibrium parameters of the reactions involving the analyte and those of any interferences present. Differences in the rates of reaction can result in even further selectivity. As an illustration, from an equilibrium standpoint, $Ni^{2+}$ and $Zn^{2+}$ should both be extractable by dithizone at the same pH. Since it takes about 2 days for the equilibrium to be reached for $Ni^{2+}$, however, $Zn^{2+}$ can be easily separated by using a 2 - 3 minute equilibration period.

Chemical kinetics not only provides means of analytical determination for a wide variety of materials, particularly interesting at trace levels, but provides a major tool for the study of chemical mechanisms, the detailed account of how reactions take place.

The guiding principle describing reaction kinetics is the Law of Mass Action which states that "the velocity of a reaction is proportional to the active concentrations of the reacting substances each raised to the power of its coefficient in the balanced equation" was clearly enunciated over a century ago (1879). Since this is introduced to students as early as 'high school chemistry', why weren't kinetic methods of analysis widely used until relatively recently? Measuring reaction velocity, which is defined as the change in the concentration of reacting substance per unit of time, requires measuring concentration as a function of time by relatively rapid as well as reliable methods which do not significantly disturb the reacting system. For reactions taking place in solution this generally means relying on an instrumental method, most notably, spectrophotometry or potentiometry.

Hence, the use of kinetic methods was limited before the development of such instrumentation. By then analysts had to 'rediscover' the power of kinetic methods. A full flowering of kinetic methods occurred in the 1960s coinciding with the development of automated instrumentation which not only rendered such methods more convenient but more reproducible as well.

In this chapter, fundamental principles and calculations of chemical kinetics and its application to analytical chemistry are described. We will try to demonstrate that the use of spreadsheet techniques makes a dramatic difference in the convenience and reliability of currently used kinetic procedures and, moreover, might lead to innovative modifications of some of these procedures. The principal advantage of the use of spreadsheets is the convenience of incorporating **all** the data in the calculation and so to ensure maximum precision and reliability.

The **rate** or **velocity** of a reaction is the change in reaction coordinate, i.e., the concentration of reactant or product, per unit of time. The reaction rate, v, which can be expressed either as the rate of disappearance of the reactant or that of the formation of the product is:

$$v = -\frac{d\,[\text{Reactant}]}{dt} = +\frac{d\,[\text{Product}]}{dt} \qquad (18\text{-}1)$$

Measurement of the reaction rate requires that the concentration of a reactant or product be measured as a function of time at frequent time intervals or, preferably, continuously with time.

In general, in a reaction in solution between substances A and B to form C, the rate expression from the Law of Mass Action can be written as

$$v = k[A]^a[B]^b \qquad (18\text{-}2)$$

where **k**, is called the **rate constant**, and the sum of the exponents, a + b, is termed the **reaction order**. The rate expression is a differential equation which can be "solved", i.e., transformed into its integrated form by conventional calculus techniques.

In this chapter, we will consider mainly first order equations because most cases of interest to us here can be configured to conform to first order kinetics, i.e., experimental conditions can be modified so that first order can be mimicked. Such reactions are described as having **pseudo first order** kinetics. This is accomplished readily by (a) keeping [B] constant. For example, if [H$^+$] or [OH$^-$] is one of the reactants, as in hydrolysis of organic esters or halides, then keeping the pH constant with a potentiostat can be used. (b) keeping [B] virtually constant, by having B in such large excess over A that, even when reaction is complete the final [B] will be about the same as the initial value.

Nuclear reactions of radioactive nuclides provide us with examples of truly **first order** reactions because the number of nuclei of a particular isotope that decay in a given time is proportional to that number. The rate expression is:

$$- \frac{dN}{dt} - kN$$

Rearranging to $\quad - \frac{dN}{N} - kdt \qquad (18\text{-}3)$

which, on integration, is

$$N - N_o e^{-kt}$$

Here, as in all first rate kinetics, a plot of the natural logarithm of the count rate vs. the time results in a straight line whose slope is proportional to the rate constant and whose intercept is $\ln N_o$. Another reaction rate characteristic, called the **half-life, $t_{\frac{1}{2}}$**. is the time required for the initial reactant concentration to be reduced to one half. For first order reactions, the $t_{\frac{1}{2}}$, independent of concentration, $= \mathbf{0.693/k}$.

If we have a mixture of isotopes with different rate constants, then measuring the total activity as a function of time will inform us about the respective amounts present at the start of the experiment and the respective rate constants as well.

## Example 18.1

Plot the radioactive decay curve of a mixture of 0.100 mc (a millicurie is equivalent to $3.700 \times 10^7$ decompositions per second(dps)) of $Ni^{65}$($t_{\frac{1}{2}}$ = 2.56hr), 0.003 mc of $Cu^{64}$($t_{\frac{1}{2}}$ = 12.8hr), and 0.002 mc of $Co^{58}$($t_{\frac{1}{2}}$ = 71d).

The number of counts after a time, t, is:

$$N - \Sigma N_o e^{-0.693 t/t^{1/2}}$$

$$N - 3.7 \times 10^6 \cdot e^{-0.693 t/2.56} + 3 \times 3.7 \times 10^6 \cdot e^{-0.693 t/12.8} + 2 \times 3.7 \times 10^6 \cdot e^{-0.693 t/(71 \times 24)}$$

$$(18\text{-}4)$$

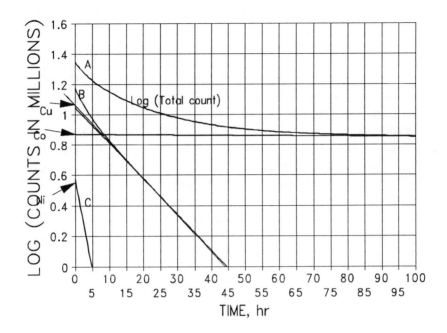

**Figure 18.1  First Order Kinetic Plot for Mixture of Radioisotopes**

Set up a spreadsheet in which column **A** is filled with values of time, in increments of 0.5 hr from 0 to 100hr. In column **B**, write the formula for N, and in **C**, the values for log N. At long times, the curve A in Figure 18.1 becomes linear because the shorter-lived isotopes have decayed away. Hence, a tangent to A at long times can be extrapolated to t = 0, to give the log $N_{Co}^{\circ}$. Such a tangent line is obtained by linear regression of the points representing counts at times > 3 $t_{1/2}$ of Co(142d < t < 213d). The resulting equation can be written as log N (or Y) = slope (or X-coefficient) ×t (or X) + log $N_o$(or constant). Hence, the slope of the regression line divided by log 2 gives the $t_{1/2}$ of Co and the intercept (X-coefficient) log $N_o$. These counting times are rather excessive, but can be avoided by using the literature value of the $t_{1/2}$ of Co to draw the tangent. Curve B was obtained by subtracting (using the spreadsheet) $N_{Co}$ at each time from the total N at the same time. (Remember to subtract the number, not the logarithm)

Curve B represents the sum of the counts for Co and Ni. Again a tangent at the longest times($ t > $ 39hr) is found by a linear regression of the points on curve B drawn and extrapolated to $ t = 0 $ to give log $N_{Cu}°$. Finally, subtract $N_{Cu}$ from the antilog of curve B, which by linear regression will then provide the basis for Curve C. Log $N_{Ni}$ as a function of time(Curve C) is linear and results in $N_{Ni}°$ at the $ t = 0 $ intercept.

This simulated determination of a ternary mixture by kinetic means is of general applicability to all systems involving concurrent first order reactions. In fact, the approach we used in Chapter 13 for spectropho-tometric analysis of a mixture of chromophoric substances by means of a multiple regression technique with the help of the spreadsheet is directly applicable to the problem of kinetic analysis of a mixture of similar substances involving a reaction that can be run under pseudo first order conditions.

## Example 18.2

Suppose we have a mixture of two esters, A and B, whose hydrolysis reaction is being run under pseudo first order conditions. The reaction is being monitored by the consumption of hydroxide ion, which gives the loss in total ester concentration as a function of time. Let the initial concentra-tions in the mixture be $C_A = 0.025$ M and $C_B = 0.043$ M, $k_A' = 10$ (in arbitrary time units) and $k_A/k_B = 10$, so that $k_B = 1$. Can the initial concentrations of A and B be computed from the total concentration measurements?

Our strategy will be to employ linear regression using $\exp(-k_A t)$ and $\exp(-k_B t)$ as independent variables just as in the multicomponent spectropho-tometric analysis the molar absorbances of each of the components were used.

Fill column **A** starting at A2 with values of time, from 0 to 10, in 0.1 increments. In column **B**, calculate $\exp(-k_A t)$[@exp(-10*A2]; in column C, $\exp(-k_B t)$[@exp(-1*A2]. Columns **D** and **E** will have $C_A$ and $C_B$ obtained by multiplying values in **B** and **C** by 0.0250 and 0.0430, respectively. Column **F** will represent the observed total ester concentration, the sum of **D2** and **E2**. Conduct linear regression by the command /**TAR**{/AR;/DAR}. For the independent variable(s) take **B2..C101**. As dependent variable, use

**F2..F101**. The resulting regression analysis yields X coefficients which are 0.0250 and 0.0430 with an intercept of essentially zero! After the first pleasant shock at such a result, remember that this simulation was carried out without taking into account any measurement error. In column **G**, multiply the total concentration, **F2**, by the noise function (see Chapter 12) using a 1%( P = 0.01) error. Repeat for P = 0.05 and 0.10. Notice that at P = 0.05, the X coefficients are 0.0265 and 0.0433, errors of 4% and < 1%, respectively. Try the entire exercise again with the $k_A/k_B$ ratio of 5 rather than 10. This time, for noise level P = 0.05, the errors in A and B rise to 20% and >6%.

## The Effect of Temperature on Reaction Rates

Chemists could not help but recognize early on that reactions vary in the time required to occur as well as their extent and wondering about the underlying reasons. If you wonder why most chemical reactions do not take place instantaneously, you probably would quickly reach the conclusion that the 'resistance' or barrier to reaction results from an energy shortage. We have been taught, after all, that a reaction will occur when its associated energy (enthalpy) change is negative (exothermic reaction). We also know that the enthalpy change is the net change involving the release of enthalpy when new bonds are formed and the loss of enthalpy when existing bonds are broken. When we consider the reaction of AB with C to form AC and B, for example, then the enthalpy change is the difference in enthalpy between the AC and AB bonds. Even if this reaction is exothermic, energy is required to break the AB bond. One way to supply this energy is to raise the temperature. Experience tells us that this usually increases the reaction rate.    The temperature dependence of the rate constant leads to the **activation energy, $E_A$,** is obtainable from exactly as the standard enthalpy,

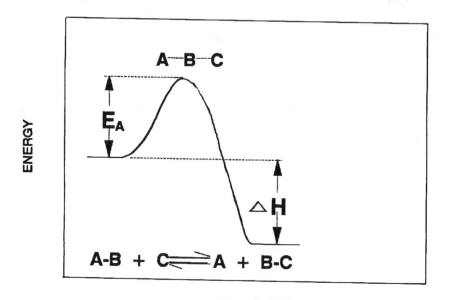

REACTION COORDINATE

**Figure 18.2   Reaction Scheme Showing Relation Between Heat of Reaction and Activation Energy**

of a reaction is obtained from the temperature dependence of the equilibrium constant, i.e.,

$$k = Ae^{-\frac{E_A}{RT}}$$

$$\text{but as } \Delta G_A = -RT\text{Ln } k = E_A - TS_A \cdots \cdots (18\text{-}5)$$

$$k = e^{\frac{\Delta S_A}{R}} e^{-\frac{E_A}{RT}}$$

This is not a coincidence. In terms of the activated complex theory, developed by the physical chemist, Henry Eyring, the reactants combine to form a complex in the "activated state", which is in simultaneous equilibrium with both reactants and products. The height of the energy barrier to the activated state complex is called the activation energy (Figure 18.2). From thermodynamic considerations alone, one cannot predict the way rate constants change from the manner in which changes in equilibrium constants occur. As seen in Figure 18.2 where the energy difference between products and reactants, $\Delta H°$, is quite different than that between reactants and activated state complex, $E_A$. In comparing families of reactions, however, frequently "extrathermodynamic" relationships, e.g., Hammett equations, between equilibrium and kinetic parameters, have been found which enable one to predict the behavior of a series of substituted reactants from that of the parent compound. This is quite useful when either the equilibrium or rate constant data is incomplete.

## Reversible Reactions

Many reactions are reversible and yet can be treated as if they go to completion provide that the extent of reaction at equilibrium is large enough, say $\sim 90\text{-}95\%$, to ignore the reverse reaction. Even if this is not the case, useful information can be obtained from the concentration - time data by using a modified equation

$$\ln \frac{C - C_E}{C_o - C_E} = (k_f - k_r)t \qquad (18\text{-}6)$$

where C is the concentration at time t, $C_o$ initial concentration, and $C_E$ the concentration at equilibrium(i.e. at a time long enough for no further changes are observed). The values of $k_f$ and $k_r$ are the (pseudo) first order rate constants for the forward and reverse reactions, respectively.

## Catalytic Reactions

A useful application of kinetic methods to trace analysis is the utilization of the catalytic properties of appropriate analytes. A color-forming reaction of an element that results in a highly colored complex provides the basis for a determination of very small quantities indeed, but ultimately the strength of the signal is limited by the low level of element present. If the element happens to act as a catalyst, however, regulating a reaction of two substances at sufficiently high concentrations, a strong signal measured as a function of time can be the basis of a much more sensitive assay.

In fact, the earliest application of kinetic methods was to determine trace levels of substances exerting catalytic activity in oxidation-reduction reactions involving multiple electron transfers (1885-trace level V on its catalysis of the oxidation of aniline). For example, the reduced form of many triphenylmethane dyes is colorless[1], and loses two electrons on oxidation to the dye. The rate of reaction with such oxidants as $IO_4^-$ is relatively slow, but can be catalyzed by trace levels of transition metal ions which involve single electron transfer in their own redox steps. Thus, trace levels of manganese can be determined by the proportionality of the rate of oxidation of leuco-malachite green by iodate at less than micromolar concentrations. Similarly, trace levels of $Cu^{2+}$, $< 10^{-6}$ M, can be determined from the catalytic effect on the atmospheric oxidation of ascorbic acid.

Such systems can be written as a generalized redox reaction
(a) The uncatalyzed reaction

$$Red + Ox \longrightarrow Products$$

(b) The catalyzed reaction, catalyst-reagent complex formation, C•Ox, in a rapid equilibrium, then a slow, rate - determining decomposition of the complex

i.                     $$C + Ox \rightleftharpoons C\bullet Ox \; ; \; \beta = [C\bullet Ox]/[C][Ox]$$

ii.                    $$C\bullet Ox + Red \rightleftharpoons Products$$

The rate expression is:

$$v = k_c[C\bullet Ox][R]$$

where $k_c$ is the rate constant of the catalyzed reaction, a value much greater than that of the uncatalyzed reaction. The [C•Ox] can be obtained from the equilibrium expression in step i. of the scheme above.

Using the familiar $\alpha$ technique for expressing concentrations,

$$[C\bullet Ox] = \alpha_{C\bullet Ox}C_T = C_T \, \beta[C]/(1 + \beta[C])$$

since

$$\alpha_{C\bullet Ox}, \; [C\bullet Ox]/C_T, \; is \; \beta[C]/(1 + \beta[C])$$

---

[1]For this reason, the reduced form is referred to as the "leuco base".

giving us finally $v = k_c C_T[R]\beta[C]/(1 + \beta[C])$ or $\approx k_c C_T[R]$ if $\beta[C] >> 1$.

This equation demonstrates the proportionality of the rate to the catalyst concentration. We have only to keep the concentration of the reducing agent constant to have our usual pseudo first order kinetics. As a corollary, if the catalyst concentration is kept constant, then the concentration of the reducing agent may be determined.

## Enzyme Catalyzed Reactions

Enzymes, biochemical catalysts, are proteins or other large molecules which have an extraordinary rate - enhancing effect on most of the reactions involved in the metabolic processes of living organisms.  Reactions such as esterification and hydrolysis, to name just two of thousands, which are exceedingly slow in the physiological pH range, are speeded up by many orders of magnitude by the appropriate enzyme.  This highly selective and active family of catalysts have wide use for enzyme assays as well as for concentration of the substances they act on(substrates).

A typical enzyme kinetic scheme resembles that of the catalyzed reaction described above, with the symbols modified.  The enzyme forms a complex with the substrate as in step i. above, which then decomposes into the products as in step ii.

$$\frac{dP}{dt} - \frac{k_c \beta[S]_T[E]_T}{1 + \beta[S]_T} \tag{18-7}$$

The simplified version of this rate, $v = k_c \beta[E]_T[S]_T$, can  be seen to serve for either enzyme or substrate concentration determinations.  If the enzyme substrate complex formation rate is not rapid, as often occurs, the rate expression that applies is somewhat more complicated and is known as the Michaelis - Menten equation

$$\frac{dP}{dt} - \frac{k_c[E]_T[S]}{[S] + (k_r + k_c)/k_f}$$

where $\dfrac{(k_r + k_c)}{k_f} - K_m$, the Michaelis – Menten constant

## PROBLEMS

**18-1**   In a solution having a 0.001 M concentration of a reactant A, how much error would be incurred in assuming pseudo first order kinetics if the concentration of B, the co-reactant is present in (a) 100- fold, (b) 50-fold, or (c) 25-fold excess of A?   The rate expression is $v = k[A][B]$.   Hint: How much change occurs in $k'$ ( $= k[B]$ and how does affect the reliability of the determination of $[A]$?

**18-2**   With an equimolar mixture of A having a $t_{1/2} = $ 90 min, and B whose $t_{1/2} = 10h$, at what time does the concentration of A become less than 5% of B, assuming that concurrent pseudo first order reactions are occurring? In what time range should concentration measurements be used to establish the tangent line?

**18-3**   Ester A and B, present in equimolar quantities, both have half-lives of 42 m at 15°C, but differ in their respective $E_a$ values, which are 5.2 Kcal/mol and 11.7 Kcal/mol, respectively. Assuming that a kinetic analysis of this mixture can be achieved when the rate constant ratio is at least 5, at what temperature will this condition be reached?

# Chapter 19

# Experimental Determination of Equilibrium Constants

The free use we have made of equilibrium constants in solving a wide variety of problems in solution chemistry is itself a most convincing argument for having reliable values of these parameters. They have a vital role in calculations designed to describe quantitatively and in detail the concentrations of all species in even the most complex, multicomponent solutions. In addition, these constants provide a valuable tool with which to relate molecular structure and chemical behavior. Correlations of equilibrium constants and structural parameters represent a useful approach for uncovering electrostatic and steric effects. Studies of such correlations have provided guidance in the design of new reagents and in the understanding of reaction pathways in metabolism, to cite only two of many applications.

The material presented throughout the text gives us a great head start. After all, measuring concentrations and using fundamental relationships to calculate K values is, in principle is almost the same as calculating concentrations from given K values that have been featured throughout the book. The variables and parameters simply change places.

This chapter will be devoted mainly to describe two experimental approaches for the determination of equilibrium constants: potentiometric and spectrophotometric. The two approaches are interrelated; a spectrophotometric method developed at the University of Arizona that provides significantly enhanced pH accuracy also impacts positively on the reliability of potentiometric pH measurements. A number of other methods including heterogeneous equilibria such as solvent extraction and ion exchange will be mentioned briefly.

Experiments designed for determination of equilibrium constant naturally include means of obtaining a significant number of replicate measurements. This can involve a series of independent single runs. It is more desirable, whenever possible, to obtain data points in the context of a titration because the medium (i.e., the solution being titrated) is changing both slowly and predictably. For this reason the needed calculational corrections and adjustments are under greater control. Also, in the time required for a careful titration many more usable data points are collected than would be the number of points obtained by batch methods. In fairness, it should be pointed out that a number of interesting systems require significantly more time to equilibrate than is usually allowed between successive titrant additions. This is a correctable problem, however.

## Potentiometric Methods

A potentiometric titration produces many individual data points that describes the titration curve in great detail. The most prominent type of such titrations uses the pH meter which electronically indicates the potential difference of the electrode pair: (1) a glass pH-indicating electrode and (2) a reference electrode which is either a calomel ($Hg_2Cl_2/Hg$) or a silver-silver chloride (AgCl/Ag) electrode. With contemporary high-input impedance pH meters, reproducibility of 0.01 to 0.05 mV, equivalent to $\pm 0.0016$ to $\pm 0.00016$ pH units can be achieved if the instrumental error were the only source.

The overall potential difference measured with the pH meter, however, is the sum of the potential differences generated in the system, and the total potentiometric error is the sum of errors in all of these differences.. This includes the one which occurs at the junction of a salt bridge (required by the calomel electrode) and the solution under test. Called the **junction potential**, its value, though generally only a few millivolts, changes with the composition of the solution being titrated. The junction potential becomes quite large at the extreme pH ranges (under 3 and over 11). Although correction is possible, it is preferable to use an electrode without liquid junction such as the AgCl/Ag electrode.

The pH meter is usually calibrated by the use of standard buffer solutions known to, at best, $\pm 0.01$ pH units. Care must be taken to control the temperature and ionic strength. A better calibration (giving corrections over a range of pH values) is obtained by using all of the pH readings in the buffer region (pH = $pK_a \pm 1$) of a titration of a standard solution of a highly characterized acid such as acetic acid (whose $pK_a$ is well known as a function of ionic strength). An additional set of calibration points can be obtained with another highly characterized system, e.g., $NH_3$ - $NH_4^+$ (with reasonable precautions, a 0.01 M $NH_3$ can be titrated with no significant $NH_3$ volatility loss). This will be mentioned again in the account of the spectrophotometric determination of equilibrium constant values.

In the titration procedure, it is the $[H^+]$ and not the $a_H$ that is needed because the equations used to find K values are stoichiometric in nature.

For this purpose, a perchloric acid solution of an ionic strength matching that of the system to be tested is titrated with NaOH and the pH meter readings in the region 3.5 to 5.0 compared with the stoichiometrically calculated pH values. The observed differences, which include the effect of activity coefficients, junction potential, and instrumental idiosyncracies, will be found to be constant ($\pm 0.01$). This difference is taken as the correction factor which converts the reading to [$H^+$]. The correction is considered applicable over the pH range from 3 to 11. This correction can also be carried out by the calibration titration technique described just above.

The titration data enable the investigator to describe the extent of reaction and therefore values of all the variables needed to calculate the K. Let us consider two cases, a simple acid-base titration for the determination of $pK_a$ values and a metal complexation reaction involving a ligand which is also a Brønsted base. The second example is based on measuring the protons that are released when the conjugate acid of the ligand reacts with a metal ion.

### Determination of $pK_a$ values of a polyprotic acid

### Example 19.1

A 50.00-mL solution of $4.944 \times 10^{-3}$ M succinic acid is titrated with 0.5110 M NaOH at 25°C and 0.1 M ionic strength. The values of pH, measured at intervals of 0.05 mL from 0 to 0.95 mL, were: 3.20, 3.40, 3.59, 3.77, 3.92, 4.07, 4.21, 4.36, 4.50, 4.65, 4.80, 4.94, 5.08, 5.22, 5.36, 5.51, 5.68, 5.88, 6.16, and 6.74.

We have already developed an equation, Equation 8-5, characterizing the titration of a weak polyprotic acid with NaOH. For our present purpose, in which pH is measured but the successive $pK_a$ values are to be determined, the calculation requires only a rearrangement of Equation 8-5. Thus,

$$\sum i\alpha_i - \frac{V_{Na}C_{Na}}{V_A C_A} - \frac{V_A + V_{Na}}{V_A C_A}([OH^-] - [H^+]) = S \qquad (19\text{-}1)$$

every term on the right hand side of 19-1 which we will call S, is known from the titration data. S is the average number of protons released from $H_N B$. A plot of S vs pH will resemble a titration curve. It can be shown that values of pH when S = 1/2, 3/2, 5/2 are equal to $pK_1$, $pK_2$, $pK_3$ provided that there is good separation between the pK values. Of course, with a monoprotic acid, at $\alpha_1 = 1/2 = \alpha_0$, pH = pK. With a diprotic acid such as succinic acid, $\alpha_1 + 2\alpha_2$ is:

$$S = \frac{K_1 H + 2K_1 K_2}{H^2 + K_1 H + K_1 K_2} \tag{19-2}$$

At S = 0.5, we may neglect $K_1 K_2$ because the second acid dissociation is repressed in the presence of the sizeable concentrations of $H_2 B$ and HB, so

$$\frac{K_1 H}{H^2 + K_1 H} = 1/2 \; ; \quad or, \quad pH = pK_1$$

At S = 1.5, a similar argument explains why the $H^2$ term may be dropped safely. Then,

$$\frac{K_1 H + 2K_1 K_2}{K_1 H + K_1 K_2} = \frac{3}{2} \quad or, \quad pH = pK_2$$

As seen from Figure 19.1, a graph of S vs. pH strongly resembles the usual acid-base titration curve, not surprising since $S = \alpha_1 + 2\alpha_2$. Values of $pK_{a1}$ and $pK_{a2}$ can be obtained at S = 0.5 and 1.5. For succinic acid, the values taken from a version of Figure 19.1 in which an expanded scale for Y (from 0.45 to 0.55 or from 1.45 to 1.55) and a corresponding expanded X scale (from 3.9 to 4.1 or from 5.3 to 5.5) are 4.018 and 5.430, respectively.

For **more precise** determination, however, use of many more points than merely those at S = 0.5, 1.5, etc. can be accomplished by rearranging Equation 19-2.

$$SH^2 = (1-S)K_1 H + (2-S)K_1 K_2$$

Divide both sides by H(1-S),

$$\frac{SH}{(1-S)} = K_1 + \frac{(2-S)}{(1-S)H} \cdot K_1 K_2 \tag{19-3}$$

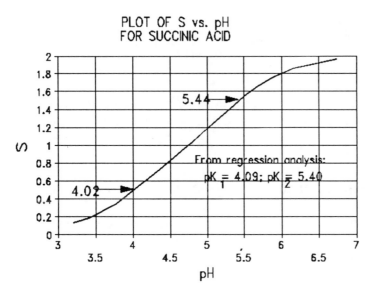

Figure 19.1 Plot of $\bar{n}_H$ vs. pH

An xy plot where SH/(1-S) is y and (2-S)/(1-S)H is x should yield a line whose slope is $K_1 K_2$ and intercept is $K_1$. The calculations are done with the spreadsheet for the titration data, using columns to tabulate $[H^+]$, $[OH^-]$, S, SH/(1-S), and (2-S)/(1-S)H so that linear regression can be carried out. Regression of the succinic acid data leads to values of the slope and intercept of Equation 19-3. The pK values obtained this way are 4.085 and 5.402. Because the pK values were so close in this case, the error resulting from using S = 0.5 and 1.5 is much greater than otherwise. Note for successive metal complex formation constants, differences can be even smaller than this case. This makes the use of equations analogous to Equation 19-3 and regression analysis mandatory.

Before considering the determination of $\beta$ values, let us take the case of the closely analogous titration of a weak polyacidic base with HCl. The equation defining such a curve is Equation 8-7. Here, instead of $\Sigma i\alpha_i$ we found $(N - \Sigma i\alpha_i)$ or $\Sigma(N-i)\alpha_i$ to be the useful function, called the **bound** protons. Together with S, the average number of free protons, they add up to N, the total available protons form the acid $H_N B$. It is more commonly referred to as $\bar{n}_H$, the average number of protons bound to the base. Our treatment of the weak acid case leading to Equation 19-3 could just as easily be written in terms of $\bar{n}_H$ which is N-S, or S'. Equation 19-3 becomes

$$\frac{(2-\bar{n}_H)H}{(\bar{n}_H-1)} - K_1 + \frac{\bar{n}_H}{(\bar{n}_H-1)H}K_1K_2 \tag{19-4}$$

The x and y variables expressed in terms of $\bar{n}_H$ are $\bar{n}_H/(\bar{n}_H-1)H$ and $(2-\bar{n}_H)H/(\bar{n}_H-1)$, respectively. The slope of this line is $K_2$ and the intercept is $K_1$.

## Determination of $\beta$ Values for Ligands that are Brønsted Bases

With proton-accepting ligands, complex formation is accompanied by proton release. Hence, these reactions can be monitored by pH measurement. Thus, for

$$M^{n+} + nHL \rightleftharpoons ML_n + nH^+$$

titrate a solution containing $C_2$ mol/l of ligand, $C_M$ of metal ion ($C_L > nC_M$), and $A$ mol/l of $HClO_4$ (to prevent complex formation prior to titration) with a standardized NaOH solution. At each NaOH addition (from which $Na$, the molar concentration of $Na^+$ in the solution is calculated), the pH is measured and, by means of a prior calibration using $HClO_4$ - NaOH titration, translated to $H$, the concentration of $H^+$ in the mixture. From this data, values of $\bar{n}$ and pL at each titration point can be obtained for use in calculating $\beta$ values.

## Method of Calculation

1.  The total number of moles H from $H_xL$ and $HClO_4$:
    Total H = $xH_xL$ + A.

2.  The available, or free H, which is = H - OH + Na

3.  Subtracting 2. from 1. gives the number of moles of H bound to the ligand:

$$\text{Bound H} = xH_xL + A - H + OH - Na \equiv S'$$

Define $\bar{n}_H$(as before) $\equiv$ average number of H bound per ligand

$$\bar{n}_H = S'/C'_L$$
$$\text{where } C'_L = C_L - \bar{n}C_M$$

$$\bar{n}_H = \frac{x[H_xL] + (x\cdots1)[H_{x-1}L] + .[HL]}{[H_xL] + [H_{x-1}L] + \cdots[L]} = x\alpha_o + (x-1)\alpha_1 + \cdots\alpha_{x-1}$$

a function solely of equilibrium constants and pH.

Similarly $\bar{n}$, the average number of ligands per metal ion,

$$\bar{n} = \frac{C_L - C'_L}{C_M} = \frac{C_L - \dfrac{S'}{\bar{n}_H}}{C_M}$$

Since both $C_L$ and $S'$ are obtained experimentally and $\bar{n}_H$ is a function of pH and previously determined pK values, $\bar{n}$, and $[L] = \alpha_L C'_L$ are defined.

Values of $\beta_i$ values may be calculated from $\bar{n}$, $[L]$ data by linearization, i.e., converting the equations to the form, $y = mx + b$, and using linear regression to get the best values. This will be illustrated with systems involving 1:1 and 1:2 complexes.

**1:1 Complex Only**

$$\bar{n} = \beta_1 L/(1 + \beta_1 L)$$

which rearranges to

$$\bar{n}/(1 - \bar{n}) = \beta_1 L$$

A plot of the left side vs. L will yield a straight line passing through the origin and having a slope of $\beta_1 L$. Linear regression offers the best value.

## 1:2 Complexes

$$\bar{n} = \frac{ML + 2ML_2}{M + ML + ML_2} = \frac{\beta_1 L + 2\beta_2 L^2}{1 + \beta_1 L + \beta_2 L^2}$$

*which rearranges to* $\dfrac{\bar{n}}{(1 - \bar{n})L} = + \dfrac{(2 - \bar{n})}{(1 - \bar{n})} L \cdot \beta_2 + \beta_1$

This function is also written in the form, $y = mx + b$, with $\beta_2$ as the slope, m, and $\beta_1$, the intercept, b.

## Example 19.2

Determination of $\beta$ Values of Cu(II) - 8-quinolinol chelates at 25°C and 0.1 M ionic strength. A solution of Cu(II) containing excess 8-quinolinol, $HClO_4$, and enough $NaClO_4$ to bring the ionic strength to 0.1 as well as 50 vol% 1,4-dioxan to avoid the formation of precipitates, was titrated with a standard solution of NaOH.

The initial solution volume was 117.9 mL, and contained a total of 0.04768 mmol of Cu, 0.2851 mmol of 8-quinolinol, and 0.4910 mmol of $HClO_4$. It was titrated with 0.0996 M NaOH. The titration data was (mL NaOH, pH): 1.40, 2.75; 1.60, 2.77; 1.80, 2.82; 2.00, 2.83; 2.20, 2.88; 2.40, 2.93; 2.60, 2.96; 2.80, 3.02; 3.00, 3.05; 3.20, 3.12; 3.40, 3.18; 3.60, 3.25; 3.80, 3.34; 4.00, 3.44.

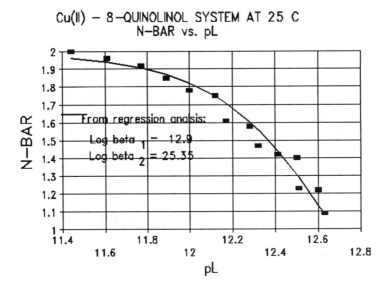

**Figure 19.2 Formation Curve for Cu(II) - 8-Quinolinol**

With the help of the spreadsheet, the appropriate calculations were carried out, using the literature values of $pK_1$ and $pK_2$ of 4.12 and 11.54, to produce Figure 19.2. Linear regression analysis of the ñ vs. pL data was conducted to yield the values shown in the figure.

## Spectrophotometric Methods

Spectrophotometric monitoring (when appropriate) of acid-base, metal complexation, or redox titrations provides a continuous recording of signals which are linearly related to the concentration of a critical species. A critical species is one whose concentration can be used, together with initial solution composition parameters, to totally define a system. In the previous section, we made use of pH as the critical species.

For this approach to be appropriate, one or more species must have a strong characteristic spectrum. With the ready availability of good spectrophotometers, species which absorb in infrared, visible, or UV regions of the spectrum are suitable. Spectrophotometers such as the increasingly common diode array for the UV and visible range and the FT/IR for the infrared range permit the rapid and reliable recording of the "entire spectrum" of each point in a titration mixture. As noted in Chapter 13, this represents a highly over determined system (one in which the number of data points and independent equations is far greater than the number of "constants" to be determined) since as many as 1000 independent measurements are contained in a single spectrum of a mixture of fewer than a dozen species. In principle, these species can therefore be determined with great precision.

Precautions must be taken to obtain reliable spectrophotometric data. A double beam instrument eliminates errors due to changes in line voltage. Furthermore, different spectrophotometric cells differ to some extent in their optical path length, requiring a correction unless the same cell is used for all solutions throughout an experiment (as with a flow-through cell). Then, too, temperature must be kept constant and measured throughout a run.

### Calculations of Spectrophotometric K Determinations

All methods of equilibrium constant determination using spectrophotometric measurements are based ultimately on the validity of Beer's Law. One of the reactants and/or products must have a characteristic spectrum from which the concentration of one of the species can be determined reliably. The concentrations of the remaining species in the equilibrated reaction mixture can then be deduced from the original solution composition and the reaction stoichiometry. Several methods will be described.

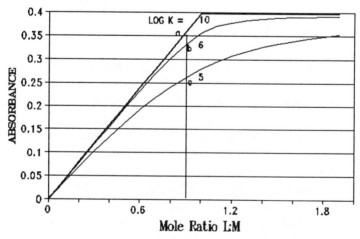

**Figure 19.3 Mole Ratio Method for $\beta_{ML}$**

**Mole-Ratio Method.**  A series of solutions containing a constant concentration of one reactant, e.g., a metal ion, but an increasing concentration of the other, e.g., a ligand. The most efficient way to conduct this experiment is as a photometric titration (See Chapter 13). The absorbance at a suitable wavelength, i.e., one at which one of the species, reactant or product, has a characteristic, strong absorbance.  In many metal complex formations, the complex itself is highly colored at wavelengths where the absorbance of the ligand does not interfere. The absorbance values are plotted against the mole ratio of ligand to metal. At very high K values, the plot will consist of two linear portions whose meeting point reveals the stoichiometry, i.e., 1:1, 2:1, etc.  At very low K values, a continuous convex curve is seen, from which the stoichiometry cannot be reliably deduced.  At each point in the figure, the difference between the absorbance of the hypothetical line (the stoichiometric line) and the experimental value is a measure of the extent of reaction, and can be used to calculate the K or $\beta$ value.

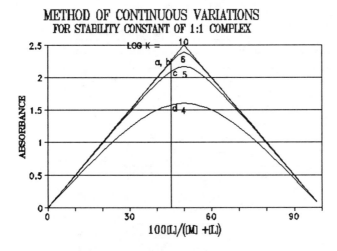

**Figure 19.4 Method of Continuous Variations**

**Method of Continuous Variations (Job's Method).**  In this method, a series of solutions are prepared in which, to take metal complexation formation as an example, the sum of the concentrations of the metal and ligand is held constant, as the mole ratio of M to L is varied from 0.0 to 1.0 in steps of 0.1 (or smaller). As with the mole ratio method, the data is seen to form two linear segments which intersect at a point that reveals the composition of the complex. For ML, this point is at mole ratio 0.5; for $ML_2$, the lines intersect at 0.33[M]:[L], etc. The differences in absorbance between the values on the hypothetical extrapolated lines and the experimental values provide the basis of calculating the $\beta$ value. A comparison of Figures 19.3 and 19.4 show that Job's method gives clearer definition of complex composition when $\beta$ values are small.

**Spectrophotometric Titration Method.** This method is no more tedious or time consuming than the other two, and has the advantage of producing data of higher quality and abundance. It represents a close analogy to the potentiometric titration method, with measurement of absorbance at a large number of points during the titration. Simultaneous measurement of pH with a glass/reference electrode pair during the titration when the pH is a relevant variable is customary. Since the volume and, usually, the ionic strength changes during titration, suitable corrections must be made.

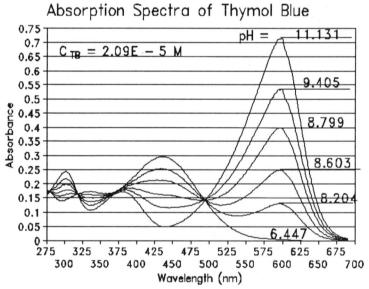

**Figure 19.5 pH Dependence of Thymol Blue Spectrum**

Consider the equilibrium involving a two-color acid-base indicator such as Thymol Blue (Figure 19.5)

$$HIn \rightleftharpoons H^+ + In; \ K_a - \frac{[H^+][In]}{[HIn]} - \frac{[H^+]\alpha_1}{\alpha_o} \qquad (13\text{-}3)$$

At pH values much higher and lower than the indicator pK the spectrum obtained for a solution of the indicator is:

$$A(f(\lambda)) = \in_{HIn}(f(\lambda))bc \text{ at pH} \leq pK$$

and

$$A(f(\lambda)) = \in_{In}(f(\lambda))bc \text{ at pH} \geq pK$$

When the pH of the solution is such that there are significant amounts of both In and HIn present, the absorbance is the sum of the contribution of the two. At any one wavelength

$$A = \in_{HIn}b[HIn] + \in_{In}b[In]$$

or

$$A/b = \in_{HIn} \alpha_o C + \in_{In}(1-\alpha_o)C$$

The value of $\alpha_o$ obtained at each of a series of wavelengths chosen so that $|\in_{HIn} - \in_{In}|$ is reasonably large [there may be one or more wavelengths where the $\in_{HIn} = \in_{In}$. These are called isosbestic points and are useful to measure C, total concentration]), is:

$$\alpha_o - \frac{A/bC - \in_{In}}{\in_{HIn} - \in_{In}} \qquad (13\text{-}4)$$

can be used to calculate the indicator acid dissociation constant since the pH is known.

$$\alpha_o - \frac{H}{H+K} \text{ or } K - \frac{H(1-\alpha_o)}{\alpha_o} \qquad (13\text{-}5)$$

By repeating the calculations at each wavelength, the error in the estimated value of K becomes quite small. Essentially all of the absorbance data can be used by using the value of $|\epsilon_{HIn} - \epsilon_{In}|^{-1}$ as a weighting factor in obtaining a best value. With spectrophotometric instruments now available in many laboratories, the residual error is largely that due to errors inherent in the pH measurement.

## PROBLEMS

**19-1**   A potentiometric titration was run for the purpose of determining $\beta_1$ and $\beta_2$ of the Ni(II)-8-quinolinol complexes. The initial solution volume was 109.7 mL, and contained a total of 0.04859 mmol of Ni, 0.4104 mmol of 8-quinolinol, and 0.4913 mmol of $HClO_4$. It was titrated with 0.1007 M NaOH. The titration data was (mL NaOH, pH):1.00, 3.20; 1.20, 3.26; 1.40, 3.32; 1.60, 3.37; 1.80, 3.42; 2.00, 3.48; 2.20, 3.53; 2.40, 3.58; 2.60, 3.63; 2.80, 3.68; 3.00, 3.73; 3.20, 3.78; 3.40, 3.84; 3.60, 3.92; 3.80, 3.97; 4.00, 4.06; 4.20, 4.13; 4.40, 4.23; 4.60, 4.33; 4.80, 4.46; 5.00, 4.60. Construct an ñ - pL curve, and use it to find the $\beta$ values (Figure 19.6). Do a linear regression to obtain the best $\beta$ values and their standard deviations.

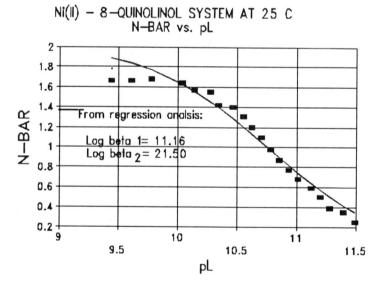

**Figure 19.6** Formation Curve for Ni(II) - 8-Quinolinol

**19-2**    A continuous variations study was carried out on mixtures of $3.0 \times 10^{-5}$ M $Fe^{3+}$ and an equimolar solution of a complexing ligand, L. Solutions were prepared by mixing from 0 to 10 mL of the $Fe^{3+}$ and $(10 - mL\ Fe^{3+})$ of L. The absorbances of each of the solutions (in sequence) were: 0.004, 0.189, 0.386, 0.584, 0.786, 0.931, 0.993, 1.059, 0.984, 0.591, and 0.003. What is the formula of this complex(molecular ratio of L/Fe)? What is its molar absorptivity? What is its formation constant and the standard deviation of $\beta$?

# APPENDIX A.   TABLE OF ATOMIC MASSES

| | | | | | |
|----|----|----------|----|----|----------|
| H | 1 | 1.0079 | Fe | 26 | 55.847 |
| He | 2 | 4.0026 | Co | 27 | 58.9332 |
| Li | 3 | 6.941 | Ni | 28 | 58.7 |
| Be | 4 | 9.01218 | Cu | 29 | 63.546 |
| B | 5 | 10.81 | Zn | 30 | 65.38 |
| C | 6 | 12.011 | Ga | 31 | 69.72 |
| N | 7 | 14.0067 | Ge | 32 | 72.59 |
| O | 8 | 15.9994 | As | 33 | 74.9216 |
| F | 9 | 18.9984 | Se | 34 | 78.96 |
| Ne | 10 | 20.179 | Br | 35 | 79.904 |
| Na | 11 | 22.98977 | Kr | 36 | 83.8 |
| Mg | 12 | 24.305 | Rb | 37 | 85.47 |
| Al | 13 | 26.98154 | Sr | 38 | 87.62 |
| Si | 14 | 28.086 | Y | 39 | 88.905 |
| P | 15 | 30.97376 | Zr | 40 | 91.22 |
| S | 16 | 32.06 | Nb | 41 | 92.906 |
| Cl | 17 | 35.453 | Mo | 42 | 95.94 |
| Ar | 18 | 39.948 | Tc | 43 | (99) |
| K | 19 | 39.098 | Ru | 44 | 101.07 |
| Ca | 20 | 40.08 | Rh | 45 | 102.905 |
| Sc | 21 | 44.9559 | Pd | 46 | 106.4 |
| Ti | 22 | 47.9 | Ag | 47 | 107.87 |
| V | 23 | 50.9414 | Cd | 48 | 112.4 |
| Cr | 24 | 51.996 | In | 49 | 114.82 |
| Mn | 25 | 54.938 | Sn | 50 | 118.69 |

| | | | | | | |
|---|---|---|---|---|---|---|
| Sb | 51 | 121.75 | Os | 76 | 190.2 |
| Te | 52 | 127.6 | Ir | 77 | 192.2 |
| I | 53 | 126.9044 | Pt | 78 | 195.09 |
| Xe | 54 | 131.3 | Au | 79 | 196.967 |
| Cs | 55 | 132.905 | Hg | 80 | 200.59 |
| Ba | 56 | 137.34 | Tl | 81 | 204.37 |
| La | 57 | 138.91 | Pb | 82 | 207.19 |
| Ce | 58 | 140.12 | Bi | 83 | 208.98 |
| Pr | 59 | 140.907 | Po | 84 | (210) |
| Nd | 60 | 144.24 | At | 85 | (210) |
| Pm | 61 | (145) | Rn | 86 | (222) |
| Sm | 62 | 150.35 | Fr | 87 | (223) |
| Eu | 63 | 151.96 | Ra | 88 | (226) |
| Gd | 64 | 157.25 | Ac | 89 | (227) |
| Tb | 65 | 158.924 | Th | 90 | 232.038 |
| Dy | 66 | 162.5 | Pa | 91 | (231) |
| Ho | 67 | 164.93 | U | 92 | 238.03 |
| Er | 68 | 167.26 | Np | 93 | (237) |
| Tm | 69 | 168.934 | Pu | 94 | (244) |
| Yb | 70 | 173.04 | Am | 95 | (243) |
| Lu | 71 | 174.97 | Cm | 96 | (245) |
| Hf | 72 | 178.49 | Bk | 97 | (247) |
| Ta | 73 | 180.948 | Cf | 98 | (249) |
| W | 74 | 183.85 | Es | 99 | (254) |
| Re | 75 | 186.2 | Fm | 100 | (252) |

# APPENDIX B. LOG $\beta$ VALUES OF AMINOACID CHELATES

| | $Mg^{2+}$ | $Ca^{2+}$ | $Mn^{2+}$ | $Fe^{2+}$ | $Co^{2+}$ | $Ni^{2+}$ | $Cu^{2+}$ | $Zn^{2+}$ | $Cd^{2+}$ |
|---|---|---|---|---|---|---|---|---|---|
| **Glycine** $pK_1=2.350$ $pK_2=9.778$ | | | | | | | | | |
| ML | 2.22 | 1.39 | 3.19 | 4.31 | 5.07 | 6.18 | 8.56 | 5.38 | 4.69 |
| $ML_2$ | | | | | 9.04 | 11.13 | 15.64 | 9.81 | 8.40 |
| $ML_3$ | | | | | 11.63 | 14.23 | | | 10.68 |
| **Alanine** $pK_1=2.348$ $pK_2=9.867$ | | | | | | | | | |
| ML | 1.96 | 1.24 | 3.02 | | 4.72 | 5.83 | 8.55 | 4.95 | |
| ML2 | | | | | 8.4 | 10.5 | 15.5 | 9.23 | |
| ML3 | | | | | 10.2 | ? | | | |
| **Aspartic Acid** $pK_3=10.002$ $pK_2=3.900$ $pK_1=1.990$ | | | | | | | | | |
| ML | 2.43 | 1.60 | 3.7 | 4.3 | 5.95 | 7.16 | 8.57 | 5.84 | 4.39 |
| ML2 | | | | | 10.2 | 12.4 | 15.4 | 10.2 | 7.55 |
| **Glutamic Acid** $pK_1=2.23$ $pK_2=4.42$ $pK_3=9.95$ | | | | | | | | | |
| ML | 2.06 | 1.17 | | 3.52 | 4.56 | 5.6 | 7.87 | 4.6 | 3.9 |
| ML2 | | | | | 7.86 | 9.8 | 14.2 | 8.3 | |
| **Tyrosine** $pK_1=2.17$ $pK_2=9.19$ $pK_3=10.47$ | | | | | | | | | |
| MHL | | 1.48 | 1.5 | | 3.87 | 5.10 | 7.81 | 4.16 | 3.57 |
| M(HL)2 | | | 5.0 | | 7.52 | 9.46 | 14.74 | 8.27 | 6.08 |
| **Cysteine** $pK_1=1.7$ $pK_2=8.36$ $pK_3=10.77$ | | | | | | | | | |
| ML | | | 4.7 | | | 9.82 | | 9.17 | |
| ML2 | | | | | | 20.07 | | 18.18 | |
| **Histidine** $pK_1=2.27$ $pK_2=6.97$ $pK_3=9.63$ | | | | | | | | | |
| ML | | | 3.3 | | 6.90 | 8.67 | 10.20 | 6.55 | 5.39 |
| $ML_2$ | | | 6.3 | | 12.34 | 15.54 | 18.10 | 12.06 | 9.66 |
| **EDMA** $pK_1=2.15$ $pK_2=6.65$ $pK_3=10.15$ | | | | | | | | | |
| ML | | | | | | 10.44 | 13.40 | 8.2 | 8.48 |
| ML2 | | | | | | 16.78 | 21.44 | | 13.23 |
| **EDDA** $pK_1=1.66$ $pK_2=2.37$ $pK_3=6.53$ $pK_4=9.59$ | | | | | | | | | |
| ML | 3.95 | | 7.05 | | 11.25 | 13.65 | 16.2 | 12.2 | 8.99 |

**EDMA** - Ethylenediamine monacetic Acid,     **EDDA** - Ethylenediamine diacetic Acid

# APPENDIX B. LOG $\beta_i$ INORGANIC COMPLEXES

| LIGANDS | Be2+ | Mg2+ | Ca2+ | Sr2+ | Ba2+ | Mn2+ | Fe2+ |
|---|---|---|---|---|---|---|---|
| HYDROXIDE | | | | | | | |
| ML | 8.6 | 2.58 | 1.3 | 0.8 | 0.6 | 3.4 | 4.5 |
| ML2 | 14.4 | | | | | | |
| ML3 | 18.8 | | | | | | |
| ML4 | 18.6 | | | | | | |
| ML5 | | | | | | | |
| ML6 | | | | | | | |
| THIOCYANATE | | | | | | | |
| ML | -0.16 | | | | | 1.23 | 1.31 |
| ML2 | -0.6 | | | | | | |
| ML3 | | | | | | | |
| ML4 | | | | | | | |
| ML5 | | | | | | | |
| ML6 | | | | | | | |
| | Be2+ | Mg2+ | Ca2+ | Sr2+ | Ba2+ | Mn2+ | Fe2+ | Co2+ |
| FLUORIDE | | | | | | | |
| ML | 4.99 | 1.8 | 1.1 | 0.1 | -0.3 | 0.7 | 0.8 |
| ML2 | 8.80 | | | | | | |
| ML3 | | | | | | | |
| ML4 | | | | | | | |
| ML5 | | | | | | | |
| ML6 | | | | | | | |
| CHLORIDE | | | | | | | |
| ML | | | | | | 0.04 | |
| ML2 | | | | | | | |
| ML3 | | | | | | | |
| ML4 | | | | | | | |
| BROMIDE | | | | | | | |
| ML | | | | | | | |
| ML2 | | | | | | | |
| ML3 | | | | | | | |
| ML4 | | | | | | | |
| ML5 | | | | | | | |
| ML6 | | | | | | | |

# APPENDIX B. LOG $\beta_i$ INORGANIC COMPLEXES

| LIGANDS | Co2+ | Ni2+ | Cu2+ | Zn2+ | Cd2+ | Hg2+ | Sn2+ | Pb2+ |
|---|---|---|---|---|---|---|---|---|
| HYDROXIDE | | | | | | | | |
| ML | 4.3 | 4.1 | 6.3 | 5.0 | 3.9 | 10.6 | | 6.3 |
| ML2 | 8.4 | 8 | | 11.1 | 7.7 | 21.8 | | 10.9 |
| ML3 | | | | 13.6 | | 20.9 | | 13.9 |
| ML4 | | | | 14.8 | 8.7 | | | |
| CYANIDE | | | | | | | | |
| ML | | | | | 6.01 | 17.00 | | |
| ML2 | | | | 11.07 | 11.12 | 32.75 | | |
| ML3 | | | | 16.05 | 15.65 | 36.31 | | |
| ML4 | | 30.22 | | 19.62 | 17.92 | 38.97 | | |
| THIOCYANATE | | | | | | | | |
| ML | -0.2 | 1.76 | 2.33 | 1.33 | 1.89 | 9.08 | 1.17k | 1.08 |
| ML2 | 1.32 | 1.58 | 3.65 | 1.91 | 2.78 | 17.26 | 1.77 | 1.48f |
| ML3 | | | | 2.0 | 2.8 | 19.97 | 1.74 | |
| ML4 | | | | 1.6 | 2.3 | 21.8 | | |
| FLUORIDE | | | | | | | | |
| ML | 0.4 | 0.5 | 1.2 | 1.15 | 0.46 | 1.6 | 4.08 | 1.44 |
| ML2 | | | | | | | 6.68 | 2.54 |
| ML3 | | | | | | | 9.5 | |
| ML4 | | | | | | | | |
| CHLORIDE | | | | | | | | |
| ML | -0.05 | -0.57 | 0.40 | 0.30 | 1.66 | 7.07 | 1.45 | 1.29f |
| ML2 | | | | 0.0 | 2.4 | 13.98 | 2.35 | 2.0f |
| ML3 | | | | 0.5 | 2.8 | 14.7 | 2.5 | 2.3f |
| ML4 | | | | 0.2 | 2.2 | 16.2 | 2.3 | 1.7f |
| BROMIDE | | | | | | | | |
| ML | -0.7 | -0.8 | -0.03 | -0.59 | 2.14 | 9.40 | 1.16 | 1.77 |
| ML2 | | | | | 3.0 | 17.98 | 1.7 | 2.6 |
| ML3 | | | | | 3.0 | 20.7 | 1.5 | 3.0 |
| ML4 | | | | | 2.9 | 22.23 | 1.0 | 2.3 |
| ML5 | | | | | | | | |
| ML6 | | | | | | | | |
| IODIDE | | | | | | | | |
| ML | | | | -1.5 | 2.28 | 12.87 | 0.70 | 1.92 |
| ML2 | | | | | 3.92 | 23.82 | 1.13 | 3.2 |
| ML3 | | | | | 5.0 | 27.6 | 2.1 | 3.9 |
| ML4 | | | | | 6.0 | 29.8 | 2.3 | 4.5 |
| ML5 | 5.13 | 8.34 | | | 6.9 | | | |
| ML6 | 4.30 | 8.31 | | | | | | |

# APPENDIX B. Log $\beta$ VALUES OF CARBOXYLIC ACID CHELATES

| | H+ | Mg2+ | Ca2+ | La3+ | UO22+ | Mn2+ | Fe2+ | Co2+ | Ni2+ |
|---|---|---|---|---|---|---|---|---|---|
| **ACETIC ACID** | | | | | | | | | |
| HL | 4.757 | | | | | | | | |
| ML | | 1.27 | 1.18 | 2.55 | 2.61 | 1.40 | 1.40 | 1.46 | 1.43 |
| ML2 | | | | | 4.9 | | | | |
| ML3 | | | | | 6.3 | | | | |
| **GLYCOLIC ACID** | | | | | | | | | |
| HL | 3.81 | | | | | | | | |
| ML | | 1.33 | 1.62 | 2.55 | 2.40 | 1.58 | | 1.97 | 2.26 |
| ML2 | | | | 4.24 | 3.96 | | | 3.01 | |
| ML3 | | | | 5.0 | 5.19 | | | | |
| **LACTIC ACID** | | | | | | | | | |
| HL | 3.860 | | | | | | | | |
| ML | | 1.37 | 1.45 | 3.3 | | 1.43 | | 1.90 | 2.22 |
| ML2 | | | | | | | | | |
| **MERCAPTOACETIC ACID** | | | | | | | | | |
| HL | 10.55 | | | | | | | | |
| H2L | 3.64 | | | | | | | | |
| MHL | | | | 1.98 | | | | | |
| ML | | | | | | 4.38 | | 5.84 | |
| ML2 | | | | | | 7.56 | 10.92 | 12.15 | 13.01 |
| ML3 | | | | | | | | | 14.99 |
| **OXALIC ACID** | | | | | | | | | |
| HL | 4.266 | | | | | | | | |
| H2L | 1.252 | | | | | | | | |
| ML | | 3.43 | 3.00 | 4.71 | 6.36 | 3.95 | | 4.72 | 5.16 |
| ML2 | | | | 7.83 | 10.59 | 4.4 | | 7.0 | |
| ML3 | | | | | | | | | |
| **PHENYLARSINODIACETIC ACID** | | | | | | | | | |
| HL | 5.03 | | | | | | | | |
| H2L | 3.61 | | | | | | | | |
| ML | | | | | | | | | 1.5 |
| ML2 | | | | | | | | | |
| **CITRIC ACID** | | | | | | | | | |
| HL | 6.396 | | | | | | | | |
| H2L | 4.761 | | | | | | | | |
| H3L | 3.128 | | | | | | | | |
| ML | | 3.37 | 4.68 | 9.18 | 7.4 | 4.15 | 4.4 | 5.00 | 5.40 |
| ML2 | | | | | 18.87 | | | | |
| M(OH)2L | | | | | | | | | |

# APPENDIX B. Log $\beta$ VALUES OF CARBOXYLIC ACID CHELATES

| | Cu2+ | Fe3+ | Ag+ | Zn2+ | Cd2+ | Hg2+ | Sn2+ | Pb2+ | Al3+ |
|---|---|---|---|---|---|---|---|---|---|
| **ACETIC ACID** | | | | | | | | | |
| HL | | | | | | | | | |
| ML | 2.22 | 3.38 | 0.73 | 1.57 | 1.93 | 5.89 | 3.3 | 2.68 | 1.51 |
| ML2 | 3.63 | 6.5 | 0.64 | 1.36 | 3.15 | | 6.0 | 4.08 | |
| ML3 | | 8.3 | | 1.57 | 2.26 | | 7.3 | | |
| **GLYCOLIC ACID** | | | | | | | | | |
| HL | | | | | | | | | |
| ML | 2.90 | 2.90 | | 2.38 | 1.87 | | | 2.23 | |
| ML2 | 4.66 | | | | | | | 3.24 | |
| ML3 | 4.27 | | | | | | | 3.2 | |
| **LACTIC ACID** | | | | | | | | | |
| HL | | | | | | | | | |
| ML | 3.02 | | | 2.22 | 1.70 | | | 2.78 | |
| ML2 | | | | 3.75 | | | | | |
| **MERCAPTOACETIC ACID** | | | | | | | | | |
| HL | | | | | | | | | |
| H2L | | | | | | | | | |
| MHL | | | | | | | | | |
| ML | | | | 7.86 | | | | | |
| ML2 | | | | 15.04 | | | | | |
| ML3 | | | | | | | | | |
| **OXALIC ACID** | | | | | | | | | |
| HL | | | | | | | | | |
| H2L | | | | | | | | | |
| ML | 6.23 | 7.74 | | | 3.89 | 9.66 | | 4.91 | 6.1 |
| ML2 | 10.27 | | | | | | | 6.76 | 11.09 |
| ML3 | | | | | | | | | 15.12 |
| **PHENYLARSINODIACETIC ACID** | | | | | | | | | |
| HL | | | | | | | | | |
| H2L | | | | | | | | | |
| ML | 2.51 | | 6.13 | 1.4 | 1.0 | 14.7 | | | |
| ML2 | | | | | | 19.92 | | | |
| **CITRIC ACID** | | | | | | | | | |
| HL | | | | | | | | | |
| H2L | | | | | | | | | |
| H3L | | | | | | | | | |
| ML | 5.90 | 11.50 | | 4.98 | 3.75 | 10.9 | | 4.34 | |
| ML2 | | | | | | | | 6.08 | |
| M(OH)2L | | | | | | | | | |

# APPENDIX B. Log $\beta$ VALUES OF CARBOXYLIC ACID CHELATES

| | H+ | Mg2+ | Ca2+ | La3+ | UO22+ | Mn2+ | Fe2+ | Co2+ | Ni2+ |
|---|---|---|---|---|---|---|---|---|---|
| **SALICYLIC ACID** | | | | | | | | | |
| HL | 13.74 | | | | | | | | |
| H2L | 2.97 | | | | | | | | |
| ML | | | | | 12.08 | 5.90 | 6.55 | 6.72 | 6.95 |
| MHL | | | | 2.08 | | | | | |
| M(HL)2 | | | | | | | | | |
| M(HL)3 | | | | | | | | | |
| M(HL)4 | | | | | | | | | |
| ML2 | | | | | 20.83 | 9.8 | 11.2 | 11.4 | 11.7 |
| ML3 | | | | | | | | | |
| **5-SULFOSALICYLIC ACID** | | | | | | | | | |
| HL | 12.53 | | | | | | | | |
| H2L | 2.84 | | | | | | | | |
| ML | | | | | 11.14 | 5.24 | 5.90 | 6.13 | 6.42 |
| ML2 | | | | | 19.20 | 8.24 | 9.9 | 9.82 | 10.24 |
| ML3 | | | | | | | | | |
| **SALICYLALDEHYDE** | | | | | | | | | |
| HL | 8.37 | | | | | | | | |
| ML | | | | 3.40 | | 2.15 | | | 3.58 |
| ML2 | | | | | | 4.0 | | | 6.5 |
| **CATECHOL** | | | | | | | | | |
| HL | 13.0 | | | | | | | | |
| H2L | 9.23 | | | | | | | | |
| ML | | 5.7p | | 9.46 | 15.9 | 7.72 | 7.95 | 8.60 | 8.92 |
| ML2 | | | | | | 13.6 | 13.5 | 15.0 | 14.4 |
| ML3 | | | | | | | | | |
| **TIRON** | | | | | | | | | |
| HL | 12.5 | | | | | | | | |
| H2L | 7.61 | | | | | | | | |
| ML | | 6.86 | 5.80 | 12.87 | 15.9 | 8.6 | | 9.49 | 9.96 |
| ML2 | | | | | | | | | |
| ML3 | | | | | | | | | |
| **PYROCATECHOL VIOLET** | | | | | | | | | |
| HL | 12.8 | | | | | | | | |
| H2L | 9.76 | | | | | | | | |
| H3L | 7.80 | | | | | | | | |
| H4L | 0.8 | | | | | | | | |
| MH2L | | | | | 7.05 | | | | |
| M(H2L)2 | | | | | 12.65 | | | | |

# APPENDIX B. Log $\beta$ VALUES OF CARBOXYLIC ACID CHELATES

| | Cu2+ | Fe3+ | Ag+ | Zn2+ | Cd2+ | Hg2+ | Sn2+ | Pb2+ | Al3+ |
|---|---|---|---|---|---|---|---|---|---|
| SALICYLIC ACID | | | | | | | | | |
| HL | | | | | | | | | |
| H2L | | | | | | | | | |
| ML | 10.62 | 16.3 | | 6.85 | 5.55 | | | | 12.9 |
| MHL | | | | | | | | | |
| M(HL)2 | | | | | | | | | |
| M(HL)3 | | | | | | | | | |
| M(HL)4 | | | | | | | | | |
| ML2 | 18.45 | 28.25 | | | | | | | 23.2 |
| ML3 | | | | | | | | | 29.8 |
| 5-SULFOSALICYLIC ACID | | | | | | | | | |
| HL | | | | | | | | | |
| H2L | | | | | | | | | |
| ML | 10.74 | 14.42 | | 6.05 | | | | | 12.3 |
| ML2 | 17.17 | 25.2 | | 10.7 | | | | | 20.0 |
| ML3 | | 32.2 | | | | | | | 25.8 |
| SALICYLALDEHYDE | | | | | | | | | |
| HL | | | | | | | | | |
| ML | 5.36 | 8.75 | | 2.87 | | | | | |
| ML2 | 10.11 | 15.55 | | 5.00 | | | | | |
| CATECHOL | | | | | | | | | |
| HL | | | | | | | | | |
| H2L | | | | | | | | | |
| ML | 13.90 | | | 9.90 | 8.2 | | | | 16.3 |
| ML2 | 24.9 | | | 9.80 | | | | | 29.3 |
| ML3 | | | | | | | | | 37.6 |
| TIRON | | | | | | | | | |
| HL | | | | | | | | | |
| H2L | | | | | | | | | |
| ML | 14.27 | 20.4 | | 10.41 | | 19.86 | | | 16.6 |
| ML2 | 25.46 | 35.4 | | 18.52 | | | | | 30.0 |
| ML3 | | 45.8 | | | | | | | 40.0 |
| PYROCATECHOL VIOLET | | | | | | | | | |
| HL | | | | | | | | | |
| H2L | | | | | | | | | |
| H3L | | | | | | | | | |
| H4L | | | | | | | | | |
| MH2L | | | | | | | | | |
| M(H2L)2 | | | | | | | | | |

# APPENDIX B. Log $\beta$ VALUES OF CARBOXYLIC ACID CHELATES

| | H+ | Mg2+ | Ca2+ | La3+ | UO22+ | Mn2+ | Fe2+ | Co2+ | Ni2+ |
|---|---|---|---|---|---|---|---|---|---|
| NITROSO-1-NAPHTHOL | | | | | | | | | |
| HL | 7.25 | | | | | | | | |
| ML | | | | | | | | | |
| 1-NITROSO-2-NAPHTHOL | | | | | | | | | |
| HL | 7.65 | | | | | | | | |
| ML | | | | | | | | | |
| ML2 | | | | | | | | | |
| NITROSO-R ACID | | | | | | | | | |
| HL | 6.88 | | | | | | | | |
| ML | | | | 4.37 | | | 2.69 | 5.40 | 6.9 |
| ML2 | | | | 7.83 | | | | | 12.5 |
| ML3 | | | | 11.24 | | | | | 17.3 |
| 4-SULFONIC ACID | | | | | | | | | |
| HL | 11.31 | | | | | | | | |
| H2L | 6.80 | | | | | | | | |
| ML | | | | | | | | 20.0 | |
| ACETYLACETONE | | | | | | | | | |
| HL | 8.99 | | | | | | | | |
| ML | | 3.65 | | 5.1s | 7.7 | 4.21 | 5.07s | 5.40 | 6.00 |
| ML2 | | 6.25 | | 9.0s | 14.1 | 7.30 | 8.67s | 9.54 | 10.60 |
| ML3 | | | | 11.9s | | | | | |
| ML4 | | | | | | | | | |
| BENZOYLACETONE | | | | | | | | | |
| HL | (8.89) | | | | | | | | |
| ML | | | | | | | | | |
| ML2 | | | | | | | | | |
| ML3 | | | | | | | | | |
| THENOYLTRIFLUOROACETONE | | | | | | | | | |
| HL | 6.53 | | | | | | | | |
| ML | | | | | | | | 3.68 | 4.45 |
| ML2 | | | | | | | | | |
| ML3 | | | | | | | | | |
| ML4 | | | | | | | | | |
| DIBENZO-18-CROWN-6-ETHER | | | | | | | | | |
| ML | | | | | | | | | |
| 2,3-DIMERCAPTOPROPANOL (BAL) | | | | | | | | | |
| HL | 10.61 | | | | | | | | |
| H2L | 8.58 | | | | | | | | |
| ML | | | | | | 5.23 | | | |
| ML2 | | | | | | 10.43 | 15.8 | | 22.8 |
| M2L3 | | | | | | | 28 | | |

# APPENDIX B. Log β VALUES OF CARBOXYLIC ACID CHELATES

| | Cu2+ | Fe3+ | Ag+ | Zn2+ | Cd2+ | Hg2+ | Sn2+ | Pb2+ | Al3+ |
|---|---|---|---|---|---|---|---|---|---|
| NITROSO-1-NAPHTHOL | | | | | | | | | |
| HL | | | | | | | | | |
| ML | | | | 3.91 | 3.33 | | | | |
| 1-NITROSO-2-NAPHTHOL | | | | | | | | | |
| HL | | | | | | | | | |
| ML | | | | 4.63 | | | | | |
| ML2 | | | | | | | | | |
| NITROSO-R ACID | | | | | | | | | |
| HL | | | | | | | | | |
| ML | 7.7 | | | 4.46 | 3.42 | | | 4.64 | |
| ML2 | 15.0 | | | 7.10 | 6.00 | | | 7.37 | |
| ML3 | | | | | | | | | |
| 4-SULFONIC ACID | | | | | | | | | |
| HL | | | | | | | | | |
| H2L | | | | | | | | | |
| ML | 21.38 | | | 12.31 | 12.74 | | | 13.19 | |
| ACETYLACETONE | | | | | | | | | |
| HL | | | | | | | | | |
| ML | 8.25 | 9.8 | | 5.06 | 3.83 | | | | 8.6 |
| ML2 | 15.05 | 18.8 | | 9.00 | 6.65 | | | | 16.5 |
| ML3 | | 26.2 | | | | | | | 22.3 |
| ML4 | | | | | | | | | |
| BENZOYLACETONE | | | | | | | | | |
| HL | | | | | | | | | |
| ML | | | | | | | | | |
| ML2 | | | | | | | | | |
| ML3 | | | | | | | | | |
| THENOYLTRIFLUOROACETONE | | | | | | | | | |
| HL | | | | | | | | | |
| ML | 5.68 | 7.18 | | | | | | | |
| ML2 | | | | | | | | | |
| ML3 | | | | | | | | | |
| ML4 | | | | | | | | | |
| DIBENZO-18-CROWN-6-ETHER | | | | | | | | | |
| ML | | | | | | | | 1.89 | |
| 2,3-DIMERCAPTOPROPANOL (BAL) | | | | | | | | | |
| HL | | | | | | | | | |
| H2L | | | | | | | | | |
| ML | | | | 13.5 | | | | | |
| ML2 | | | | 23.3 | | | | | |
| M2L3 | | | | 40.4 | | | | | |

# APPENDIX B. Log $\beta$ VALUES OF CARBOXYLIC ACID CHELATES

| | H+ | Mg2+ | Ca2+ | La3+ | UO22+ | Mn2+ | Fe2+ | Co2+ | Ni2+ |
|---|---|---|---|---|---|---|---|---|---|
| **DIMETHYLGLYOXIME** | | | | | | | | | |
| L | 11.9 | | | | | | | | |
| HL | 10.45 | | | | | | | | |
| ML2 | | | | | | | | | 17.24 |
| **THIOUREA** | | | | | | | | | |
| ML | | | | | | | | | |
| ML2 | | | | | | | | | |
| ML3 | | | | | | | | | |
| ML4 | | | | | | | | | |
| ML5 | | | | | | | | | |
| ML6 | | | | | | | | | |
| **DIPHENYLTHIOCARBAZONE** **DITHIZONE** | | | | | | | | | |
| HL | 4.45 | | | | | | | | |
| ML | | | | | | | | | |
| ML2 | | | | | | | | | |
| M.L | | | | | | | | | |
| | | | | | | | | | |

| | Cu2+ | Fe3+ | Ag+ | Zn2+ | Cd2+ | Hg2+ | Sn2+ | Pb2+ | Al3+ |
|---|---|---|---|---|---|---|---|---|---|
| **DIMETHYLGLYOXIME** | | | | | | | | | |
| L | | | | | | | | | |
| HL | | | | | | | | | |
| ML2 | 19.24 | | | | | | | | |
| **THIOUREA** | | | | | | | | | |
| ML | | | 7.11b | 0.5d | 1.5 | | | 0.63 | |
| ML2 | | | 10.61b | 0.8d | 2.12e | 22.1 | | 1.37 | |
| ML3 | | | 12.73b | 0.9d | 3.65e | 24.7 | | 1.8 | |
| ML4 | | | 13.57b | | 4.0e | 26.5 | | 2.0 | |
| ML5 | | | | | | | | 2.0 | |
| ML6 | | | | | | | | 2.0 | |
| **DIPHENYLTHIOCARBAZONE** **(DITHIZONE)** | | | | | | | | | |
| HL | | | | | | | | | |
| ML | | | | 7.75 | | | | | |
| ML2 | 22.3 | | | 15.05 | | | | 15.2 | |
| M.L | | | | | | | | -23.7 | |
| | -30.7 | | | | | | | | |

# APPENDIX B.  Log $\beta$ VALUES OF POLYAMINOCARBOXYLIC ACID CHELATES

| Ligands: | EDTA | HEDTA | DTPA | NTA | IDA |
|---|---|---|---|---|---|
| Successive pK<br>Values of<br>Ligands | 2.0<br>2.68<br>6.11<br>10.17 | 2.3<br>5.11<br>8.65 | 2.3<br>2.6<br>4.17<br>8.26<br>9.48 | 1.8<br>2.48<br>9.65 | 1.88<br>2.84<br>9.79 |
| Metal Ions | | $\log \beta_1$<br>and | $\log \beta_2$ | Values | |
| $Mg^{2+}$ | 8.83 | 7.0 | 9.34 | 5.47 | 2.98 |
| $Ca^{2+}$ | 10.61 | 8.2 | 10.75 | 6.39, 8.76 | 2.59 |
| $Ba^{2+}$ | 7.81 | 6.2 | 8.78 | 4.80 | 1.67 |
| $La^{3+}$ | 15.46 | 13.56 | 19.48 | 10.47, 17.83 | 5.88, 9.97 |
| $Ce^{3+}$ | 15.94 | 14.21 | 20.33 | 10.70, 18.66 | 6.18, 10.71 |
| $Th^{4+}$ | 23.2 | 18.5 | 28.8 | 13.3 | |
| $UO_2^{2+}$ | 7.40 | | | 9.50 | 8.96 |
| $Mn^{2+}$ | 13.81 | 10.8 | 15.51 | 7.46, 10.94 | |
| $Fe^{2+}$ | 14.27 | 12.2 | 16.4 | 8.33, 12.8 | 5.8, 10.1 |
| $Co^{2+}$ | 16.26 | 14.5 | 19.15 | 10.38, 14.33 | 6.97, 12.31 |
| $Ni^{2+}$ | 18.52 | 17.1 | 20.17 | 11.50, 16.32 | 9.24, 15.71 |
| $Cu^{2+}$ | 18.70 | 17.5 | 21.38 | 12.94, 17.42 | 10.57, 16.54 |
| $Fe^{3+}$ | 25.0 | 19.8 | 28.0 | 15.9, 24.3 | 10.72 |
| $Zr^{4+}$ | 29.4 | | 35.8 | 20.8 | |
| $Ag^+$ | 7.32 | 6.71 | 8.61 | 5.2 | |
| $Pd^{2+}$ | 18.5 | | | 19.3 | |
| $Zn^{2+}$ | 16.44 | 14.6 | 18.29 | 10.66, 14.24 | 7.24, 12.52 |
| $Cd^{2+}$ | 16.36 | 13.1 | 19.0 | 9.78, 14.39 | 5.71, 10.12 |
| $Hg^{2+}$ | 21.5 | 20.05 | 26.40 | 14.6 | 11.76 |
| $Pb^{2+}$ | 17.88 | 15.5 | 18.66 | 11.34 | 7.31 |
| $Al^{3+}$ | 16.5 | 14.4 | 18.7 | 11.4 | 8.1 |
| $Ga^{3+}$ | 20.3 | 16.9 | 25.54 | 13.6 | |
| $In^{3+}$ | 24.9 | 20.2 | 29.0 | 16.9 | |
| $Bi^{3+}$ | 27.8 | 22.3 | 35.0 | 17.5, 26.0 | |

EDTA - Ethylenedinitrilotetraacetic Acid, HEDTA - Hydroxyethylethyleneditrilotriacetic acid
DTPA - Diethylenetrinitrilopentaacetic acid, IDA - Iminodiacetic acid, NTA - Nitrilotriacetic Acid

# APPENDIX C. SOLUBILITY PRODUCT CONSTANTS AT 25°C

| Substance | Formula | $K_{sp}$ |
|-----------|---------|----------|
| Aluminum arsenate | $AlAsO_4$ | $1.6 \times 10^{-16}$ |
| Aluminum hydroxide | $Al(OH)_3$ | $2.0 \times 10^{-32}$ |
| Aluminum phosphate | $AlPO_4$ | $5.8 \times 10^{-19}$ |
| Aluminum sulfide | $Al_2S_3$ | $2.0 \times 10^{-7}$ |
| Barium arsenate | $Ba_3(AsO_4)_2$ | $1.1 \times 10^{-13}$ |
| Barium bromate | $Ba(BrO_3)_2$ | $5.5 \times 10^{-6}$ |
| Barium carbonate | $BaCO_3$ | $5.1 \times 10^{-9}$ |
| Barium chromate | $BaCrO_4$ | $1.2 \times 10^{-10}$ |
| Barium fluoride | $BaF_2$ | $1.0 \times 10^{-6}$ |
| Barium iodate | $Ba(IO_3)_2$ | $1.5 \times 10^{-9}$ |
| Barium oxalate | $BaC_2O_4$ | $1.5 \times 10^{-8}$ |
| Barium phosphate | $Ba_3(PO_4)_2$ | $6.0 \times 10^{-39}$ |
| Barium sulfate | $BaSO_4$ | $1.0 \times 10^{-10}$ |
| Cadmium arsenate | $Cd_3(AsO_4)_2$ | $2.2 \times 10^{-33}$ |
| Cadmium carbonate | $CdCO_3$ | $5.2 \times 10^{-12}$ |
| Cadmium hydroxide | $Cd(OH)_2$ | $2.8 \times 10^{-14}$ |
| Cadmium oxalate | $CdC_2O_4$ | $2.8 \times 10^{-8}$ |
| Cadmium sulfide | $CdS$ | $7.0 \times 10^{-27}$ |
| Calcium arsenate | $Ca_3(AsO_4)_2$ | $6.8 \times 10^{-19}$ |
| Calcium carbonate | $CaCO_3$ | $4.7 \times 10^{-9}$ |
| Calcium chromate | $CaCrO_4$ | $7.1 \times 10^{-4}$ |
| Calcium hydroxide | $Ca(OH)_2$ | $5.5 \times 10^{-6}$ |
| Calcium fluoride | $CaF_2$ | $4.9 \times 10^{-11}$ |
| Calcium iodate | $Ca(IO_3)_2$ | $5.5 \times 10^{-6}$ |

303

# APPENDIX C. SOLUBILITY PRODUCT CONSTANTS AT 25°C

| Substance | Formula | $K_{sp}$ |
|---|---|---|
| Calcium oxalate | $CaC_2O_4$ | $2.1 \times 10^{-9}$ |
| Calcium phosphate | $Ca_3(PO_4)_2$ | $2.0 \times 10^{-29}$ |
| Calcium sulfate | $CaSO_4$ | $2.4 \times 10^{-5}$ |
| Cerium(III) oxalate | $Ce_2(C_2O_4)_3$ | $2.5 \times 10^{-29}$ |
| Cerium(III) sulfide | $Ce_2S_3$ | $6.0 \times 10^{-11}$ |
| Chromium arsenate | $CrAsO_4$ | $7.8 \times 10^{-21}$ |
| Chromium hydroxide | $Cr(OH)_3$ | $7.0 \times 10^{-31}$ |
| Cobalt carbonate | $CoCO_3$ | $8.0 \times 10^{-13}$ |
| Cobalt hydroxide | $Co(OH)_2$ | $2.0 \times 10^{-16}$ |
| Cobalt oxalate | $CoC_2O_4$ | $4.0 \times 10^{-8}$ |
| Cobalt sulfide | $CoS$ | $8.0 \times 10^{-23}$ |
| Copper(I) bromide | $CuBr$ | $5.3 \times 10^{-9}$ |
| Copper(I) chloride | $CuCl$ | $3.2 \times 10^{-7}$ |
| Copper(I) cyanide | $CuCN$ | $1.0 \times 10^{-11}$ |
| Copper(I) iodide | $CuI$ | $1.1 \times 10^{-12}$ |
| Copper(I) sulfide | $Cu_2S$ | $1.0 \times 10^{-48}$ |
| Copper(I) thiocyanate | $CuNCS$ | $1.6 \times 10^{-11}$ |
| Copper(II) carbonate | $CuCO_3$ | $2.5 \times 10^{-10}$ |
| Copper(II) chromate | $CuCrO_4$ | $3.6 \times 10^{-6}$ |
| Copper(II) hydroxide | $Cu(OH)_2$ | $2.2 \times 10^{-20}$ |
| Copper(II) sulfide | $CuS$ | $8.0 \times 10^{-36}$ |
| Iron(II) carbonate | $FeCO_3$ | $3.5 \times 10^{-11}$ |
| Iron(II) hydroxide | $Fe(OH)_2$ | $8.0 \times 10^{-16}$ |

# APPENDIX C. SOLUBILITY PRODUCT CONSTANTS AT 25°C

| Substance | Formula | $K_{sp}$ |
|---|---|---|
| Iron(II) oxalate | $FeC_2O_4$ | $2.0 \times 10^{-7}$ |
| Iron(II) sulfide | $FeS$ | $5.0 \times 10^{-18}$ |
| Iron(III) arsenate | $FeAsO_4$ | $5.8 \times 10^{-21}$ |
| Iron(III) hydroxide | $Fe(OH)_3$ | $6.0 \times 10^{-38}$ |
| Iron(III) phosphate | $FePO_4$ | $1.3 \times 10^{-22}$ |
| Iron(III) sulfide | $Fe_2S_3$ | $1.0 \times 10^{-88}$ |
| Lead arsenate | $Pb_3(AsO_4)_2$ | $4.1 \times 10^{-36}$ |
| Lead bromide | $PbBr_2$ | $4.6 \times 10^{-6}$ |
| Lead carbonate | $PbCO_3$ | $1.5 \times 10^{-13}$ |
| Lead chloride | $PbCl_2$ | $1.6 \times 10^{-5}$ |
| Lead chromate | $PbCrO_4$ | $2.0 \times 10^{-16}$ |
| Lead fluoride | $PbF_2$ | $2.7 \times 10^{-8}$ |
| Lead formate | $Pb(CHO_2)_2$ | $2.0 \times 10^{-7}$ |
| Lead hydroxide | $Pb(OH)_2$ | $4.0 \times 10^{-15}$ |
| Lead iodate | $Pb(IO_3)_2$ | $2.6 \times 10^{-13}$ |
| Lead iodide | $PbI_2$ | $7.1 \times 10^{-9}$ |
| Lead molybdate | $PbMoO_4$ | $4.0 \times 10^{-6}$ |
| Lead oxalate | $PbC_2O_4$ | $8.0 \times 10^{-12}$ |
| Lead sulfate | $PbSO_4$ | $1.7 \times 10^{-8}$ |
| Lead sulfide | $PbS$ | $8.0 \times 10^{-28}$ |
| Lead thiocyanate | $Pb(NCS)_2$ | $3.0 \times 10^{-8}$ |
| Magnesium ammonium phosphate | $MgNH_4PO_4$ | $2.5 \times 10^{-13}$ |
| Magnesium arsenate | $Mg_3(AsO_4)_2$ | $2.1 \times 10^{-20}$ |
| Magnesium carbonate | $MgCO_3$ | $1.0 \times 10^{-5}$ |

# APPENDIX C. SOLUBILITY PRODUCT CONSTANTS AT 25°C

| Substance | Formula | $K_{sp}$ |
|---|---|---|
| Magnesium fluoride | $MgF_2$ | $6.4 \times 10^{-9}$ |
| Magnesium hydroxide | $Mg(OH)_2$ | $1.1 \times 10^{-11}$ |
| Magnesium oxalate | $MgC_2O_4$ | $8.6 \times 10^{-5}$ |
| Manganese(II) arsenate | $Mn_3(AsO_4)_2$ | $1.9 \times 10^{-29}$ |
| Manganese(II) carbonate | $MnCO_3$ | $8.8 \times 10^{-11}$ |
| Manganese(II) hydroxide | $Mn(OH)_2$ | $1.6 \times 10^{-13}$ |
| Manganese(II) oxalate | $MnC_2O_4$ | $1.0 \times 10^{-15}$ |
| Manganese(II) sulfide | $MnS$ | $1.0 \times 10^{-11}$ |
| Mercury(I) acetate | $Hg_2(C_2H_3O_2)_2$ | $3.5 \times 10^{-10}$ |
| Mercury(I) bromide | $Hg_2Br_2$ | $1.3 \times 10^{-22}$ |
| Mercury(I) carbonate | $Hg_2CO_3$ | $8.9 \times 10^{-17}$ |
| Mercury(I) chloride | $Hg_2Cl_2$ | $1.3 \times 10^{-18}$ |
| Mercury(I) oxide | $Hg_2O$ | $1.6 \times 10^{-23}$ |
| Mercury(I) iodate | $Hg_2(IO_3)_2$ | $2.5 \times 10^{-14}$ |
| Mercury(I) chromate | $Hg_2CrO_4$ | $2.0 \times 10^{-9}$ |
| Mercury(I) iodide | $Hg_2I_2$ | $4.0 \times 10^{-29}$ |
| Mercury(I) sulfate | $Hg_2SO_4$ | $6.8 \times 10^{-7}$ |
| Mercury(II) bromide | $HgBr_2$ | $1.1 \times 10^{-19}$ |
| Mercury(II) chloride | $HgCl_2$ | $6.1 \times 10^{-15}$ |
| Mercury(II) iodate | $Hg(IO_3)_2$ | $3.0 \times 10^{-13}$ |
| Mercury(II) iodide | $HgI_2$ | $4.0 \times 10^{-29}$ |
| Mercury(II) oxide | $HgO$ | $3.0 \times 10^{-26}$ |
| Mercury(II) sulfide | $HgS$ (black) | $3.0 \times 10^{-52}$ |

# APPENDIX C.  SOLUBILITY PRODUCT CONSTANTS AT 25°C

| Substance | Formula | $K_{sp}$ |
|---|---|---|
| Nickel arsenate | $Ni_3(AsO_4)_2$ | $3.1 \times 10^{-26}$ |
| Nickel hydroxide | $Ni(OH)_2$ | $2.0 \times 10^{-15}$ |
| Nickel iodate | $Ni(IO_3)_2$ | $1.4 \times 10^{-8}$ |
| Nickel oxalate | $NiC_2O_4$ | $4.0 \times 10^{-10}$ |
| Nickel sulfide | $NiS$ | $2.0 \times 10^{-21}$ |
| Silver acetate | $AgC_2H_3O_2$ | $2.3 \times 10^{-3}$ |
| Silver arsenate | $Ag_3AsO_4$ | $1.0 \times 10^{-22}$ |
| Silver bromide | $AgBr$ | $5.2 \times 10^{-13}$ |
| Silver carbonate | $Ag_2CO_3$ | $8.2 \times 10^{-12}$ |
| Silver chloride | $AgCl$ | $1.0 \times 10^{-10}$ |
| Silver chromate | $Ag_2CrO_4$ | $2.4 \times 10^{-12}$ |
| Silver cyanide | $AgCN$ | $2.0 \times 10^{-16}$ |
| Silver iodate | $AgIO_3$ | $3.1 \times 10^{-8}$ |
| Silver iodide | $AgI$ | $8.3 \times 10^{-17}$ |
| Silver molybdate | $Ag_2MoO_4$ | $2.8 \times 10^{-12}$ |
| Silver oxalate | $Ag_2C_2O_4$ | $1.0 \times 10^{-11}$ |
| Silver oxide | $Ag_2O$ | $2.6 \times 10^{-8}$ |
| Silver phosphate | $Ag_3PO_4$ | $1.0 \times 10^{-21}$ |
| Silver sulfate | $Ag_2SO_4$ | $1.7 \times 10^{-5}$ |
| Silver sulfide | $Ag_2S$ | $7.0 \times 10^{-50}$ |
| Silver thiocyanate | $AgNCS$ | $1.0 \times 10^{-12}$ |
| Strontium carbonate | $SrCO_3$ | $7.0 \times 10^{-10}$ |
| Strontium chromate | $SrCrO_4$ | $5.0 \times 10^{-6}$ |

# APPENDIX C. SOLUBILITY PRODUCT CONSTANTS AT 25°C

| Substance | Formula | $K_{sp}$ |
|---|---|---|
| Strontium fluoride | $SrF_2$ | $7.9 \times 10^{-10}$ |
| Strontium iodate | $Sr(IO_3)_2$ | $3.3 \times 10^{-7}$ |
| Strontium oxalate | $SrC_2O_4$ | $5.6 \times 10^{-8}$ |
| Strontium phosphate | $Sr_3(PO_4)_2$ | $1.0 \times 10^{-31}$ |
| Strontium sulfate | $SrSO_4$ | $7.6 \times 10^{-7}$ |
| Thallium(I) bromide | $TlBr$ | $3.9 \times 10^{-6}$ |
| Thallium(I) chloride | $TlCl$ | $1.8 \times 10^{-4}$ |
| Thallium(I) chromate | $l_2CrO_4$ | $9.8 \times 10^{-13}$ |
| Thallium(I) iodide | $TlI$ | $6.5 \times 10^{-8}$ |
| Thallium(I) phosphate | $Tl_3PO_4$ | $6.7 \times 10^{-8}$ |
| Tin(II) hydroxide | $Sn(OH)_2$ | $1.6 \times 10^{-27}$ |
| Tin(II) sulfide | $SnS$ | $1.3 \times 10^{-27}$ |
| Tin(IV) hydroxide | $Sn(OH)_4$ | $1.0 \times 10^{-57}$ |
| Zinc arsenate | $Zn_3(AsO_4)_2$ | $1.1 \times 10^{-27}$ |
| Zinc carbonate | $ZnCO_3$ | $2.1 \times 10^{-11}$ |
| Zinc hydroxide | $Zn(OH)_2$ | $7.0 \times 10^{-18}$ |
| Zinc iodate | $Zn(IO_3)_2$ | $2.0 \times 10^{-8}$ |
| Zinc oxalate | $ZnC_2O_4$ | $2.5 \times 10^{-9}$ |
| Zinc sulfide (ß form) | $ZnS$ | $3.0 \times 10^{-22}$ |

# Index